D0503446

Mathematics Workbook

AGE 9–11

10-Minute Maths Tests

David E Hanson

GALORE PARK

AN HACHETTE UK COMPANY

About the author

David Hanson has over 40 years' experience of teaching and has been leader of the Independent Schools Examinations Board (ISEB) 11+ Maths setting team, a member of the ISEB 13+ Maths setting team and a member of the ISEB Editorial Endorsement Committee.

Every effort has been made to trace all copyright holders, but if any have been inadvertently overlooked the publishers will be pleased to make the necessary arrangements at the first opportunity.

Although every effort has been made to ensure that website addresses are correct at time of going to press, Galore Park cannot be held responsible for the content of any website mentioned in this book. It is sometimes possible to find a relocated web page by typing in the address of the home page for a website in the URL window of your browser.

Hachette UK's policy is to use papers that are natural, renewable and recyclable products and made from wood grown in sustainable forests. The logging and manufacturing processes are expected to conform to the environmental regulations of the country of origin.

Orders: please contact Bookpoint Ltd, 130 Milton Park, Abingdon, Oxon OX14 4SB. Telephone: +44 (0)1235 827827. Lines are open 9.00a.m.–5.00p.m., Monday to Saturday, with a 24-hour message answering service. Visit our website at www.galorepark.co.uk for details of other revision guides for Common Entrance, examination papers and Galore Park publications.

Published by Galore Park Publishing Ltd
An Hachette UK company
Carmelite House, 50 Victoria Embankment,
London EC4Y 0DZ

Text copyright © David Hanson 2014
The right of David Hanson to be identified as the author of this Work has been asserted by him in accordance with sections 77 and 78 of the Copyright, Designs and Patents Act 1988.

Impression number 10 9 8 7 6
2019

All rights reserved. No part of this publication may be sold, reproduced, stored in a retrieval system, or transmitted, in any form or by any means, electronic, mechanical, photocopying, recording, or otherwise, without either the prior written permission of the copyright owner or a licence permitting restricted copying issued by the Copyright Licensing Agency, Saffron House, 6–10 Kirby Street, London EC1N 8TS.

Typeset in India
Printed in India
Illustrations by Aptara, Inc.

A catalogue record for this title is available from the British Library.

ISBN: 978 1 471829 63 5

Contents and test results

Answers to all the questions in this book can be found in the pull-out section in the middle.

Introduction

The material in this workbook takes account of recent developments in the National Curriculum but, for convenience, the earlier 'strand' divisions are used as a basis for grouping the topics.

The book consists of 70 tests. There are 10 tests in each of the six subject 'strands' and 10 mixed tests:

Strands in this book	Topics covered	New National Curriculum
Number	Properties of numbers: multiples and factors, counting backwards and forwards, sequences, prime numbers, negative numbers, squares and cubes	Number (1) – number and place value with some elements of Number (4)
	Place value and ordering: words and numerals, Roman numerals, place value including abacus, multiplying by powers of 10	
	Estimation and approximation: estimating position on a number line, rounding to the nearest power of 10, nearest integer and 1 decimal place	
	Fractions, decimals and percentages: equivalent fractions, ordering fractions, adding and subtracting fractions, multiplying and dividing fractions, equivalence of fractions, decimals and percentages, fractions and percentages of quantities	Number (4) – fractions (including decimals and percentages)
Calculations	Number operations: understanding and using basic number facts, order of operations including use of brackets	Number (2) – addition and subtraction
	Mental strategies: practice using a variety of strategies, including making use of known facts	Number (3) – multiplication and division
	Written methods: practice in all methods with integers and decimals	
	Interpreting and checking results: practice using even and odd number facts, and approximations	
Problem solving	Decision making and reasoning about numbers or shapes: looking for patterns, sequences	Number (1)
		Number (2)
	Real-life mathematics: applying mathematical skills to everyday problems	Number (3)
		Number (4)
Pre-algebra	Missing numbers in number sentences, simple word formulae, sequences, number machines	Number (2)
		Number (3)
Shape, space and measures	Measures: measurement units and conversions, reading scales, area and perimeter of rectilinear shapes, time	Measurement
	Shape: names and symmetry of plane shapes, polygons, names and nets of solid shapes	Geometry (1) – properties of shapes
	Space: names of angles, calculating angles in a right angle and at a straight line, equal sides and angles of plane shapes, points on a co-ordinate grid, translation of a shape on a co-ordinate grid	Geometry (2) – position and direction
Handling data	Data tables, pictograms, bar charts, tallies, frequency diagrams, simple Carroll diagrams and Venn diagrams, line graphs, likelihood	Statistics
Mixed tests	A selection of questions from all six of the strands above.	

The strand tests:

- consist of 10 questions printed in two columns on one page
- can be tackled in a suggested time of 10 minutes
- contain questions on different topics of the 'strand'
- cover the key ideas in the strand
- feature a gradual increase in difficulty through the ten tests of each strand
- follow the same general pattern, and are designed to:
  - build confidence
  - facilitate the identification of weak areas
  - provide practice in recalling facts and procedures
  - facilitate the monitoring of progress
  - encourage working quickly and accurately.

Two consecutive strand tests – for example 1 and 2, 3 and 4 or 5 and 6 – could be combined to form an assessment covering the whole strand, to be tackled in a time of 20 minutes.

The mixed tests consist of 10 questions from different sections of the strands.

Two consecutive mixed tests – for example 1 and 2, 3 and 4 or 5 and 6 – could be combined to form an assessment covering the whole curriculum, to be tackled in a time of 20 minutes.

Using the tests

The tests can be used in two main ways:

- Complete the test, as quickly as possible, recording the time taken.
- Do as much as possible in a fixed time.

Responses should be written or drawn on the page but additional paper should be available to do extra working if required.

Answers

Answers to the questions can be pulled out of the middle of the book.

Where appropriate, answers involving fractions should be given in their simplest form.

Marks

Each question has a total of two marks. The number of parts in a question varies and it is left to the marker to decide on allocation of the 2 marks. For a question with an odd number of parts, give half marks to the easier parts and whole marks to the more difficult parts. It is left to the discretion of markers to award a half mark if this is considered appropriate, particularly in the early stages or for a weak pupil.

Notes

In **factor rainbows**, it helps to link factor pairs with arcs as shown in this example:

1 2 3 4 6 9 12 18 36

Number test 1

1 (a) On the number grid, shade all the multiples of 3 and circle all the multiples of 9

22	23	24	25	26	27
32	33	34	35	36	37
42	43	44	45	46	47
52	53	54	55	56	57

Accuracy
Multiples

Accuracy

(b) Write down the number between 20 and 30 which is a multiple of both 3 and 7

21

2 (a) What is the highest common factor of 12 and 16?

4

(b) Complete the factor rainbow for 18

1 __2__ __3__ __6__ __9__ 18

3 Write the next two terms in each sequence.

(a) 1, 4, 7, 10, __13__ , __16__

(b) 1, 3, 9, 27, __81__ , __243__

4 (a) List the prime numbers between 10 and 20 __11, 13, 17, 19__

(b) Write 24 as a product of its prime factors.

3 × 2 × 2 × 2 = 24

5 (a) What temperature is 5 degrees lower than 1 °C? __ -4 °C__

(b) What is the value of -5 + 6? __1__

6 Write down

(a) the square of 4 ___2___   ×16

(b) the square root of 9 ___3___ ✓

(c) the cube of 2 ___8___

7 (a) Write fifteen thousand, two hundred and four in figures.

15,204

(b) Write 12 307 in words.

twelve thousand three hundred and seven

8 (a) Multiply 5175 by 10 51750

(b) Divide 407 by 100 4.07

9 (a) If these decimals are arranged in order of size, which will be in the middle?

6.75

7.65   7.56   6.57   6.75   5.76

(b) Write down the number which is exactly half way between 10 and 26

18

10 (a) Circle and label the positions 7 and -3

-5   -3   0   5   7

(b) Circle the numbers n such that $n < 4$

-5   0   5

24

3   8

2   4

2   2

7

Number test 2

1 (a) Mark and label the position of 5

5

0 ——————————————— 10

(b) Mark and label the position of 75

75

0 ——————————————— 100

2 (a) Round 53 678 to the nearest 1000

54 000

(b) Round 508 058 to the nearest 100

508 100

3 (a) Round 3.409 to 2 decimal places.

3.41

(b) Round 3.45 to 1 decimal place.

3.5

4 (a) Round 10.9 to 2 significant figures.

11

(b) Round 4.45 to 1 significant figure.

4

5 (a) Complete these equivalent fractions:

$\frac{1}{3} = \frac{4}{12}$

(b) Write the fraction $\frac{12}{20}$ in its simplest

form (lowest terms).

$\frac{3}{5}$

(c) Write the mixed fraction $1\frac{1}{2}$ as an

improper fraction.

$\frac{3}{2}$

6 (a) Add: $\frac{3}{4} + \frac{1}{2}$

$\frac{5}{4} = 1$

(b) Subtract: $\frac{2}{3} - \frac{1}{2}$

$\frac{0.5}{3}$ or

7 (a) Multiply: $\frac{1}{2} \times \frac{2}{3}$

$\frac{2}{6}$ or $\frac{1}{3}$

(b) Divide: $\frac{3}{4} \div \frac{1}{2}$

$\frac{6}{4}$ or

8 (a) Write the fraction $\frac{1}{5}$ as a decimal.

0.20

(b) Write the decimal 0.25 as a fraction in its simplest form.

$\frac{2.5}{4}$

(c) Write the fraction $\frac{3}{10}$ as a percentage.

30 %

9 (a) What is 20% of 60 kg? _____ 12 kg

(b) What is $\frac{1}{4}$ of 72 cm? _____ 18 cm

10 (a) If Bill and Mary share 20 stamps in the ratio 1:4, how many stamps will Bill get?

4

(b) What proportion of this strip is shaded?

$\frac{3}{8}$

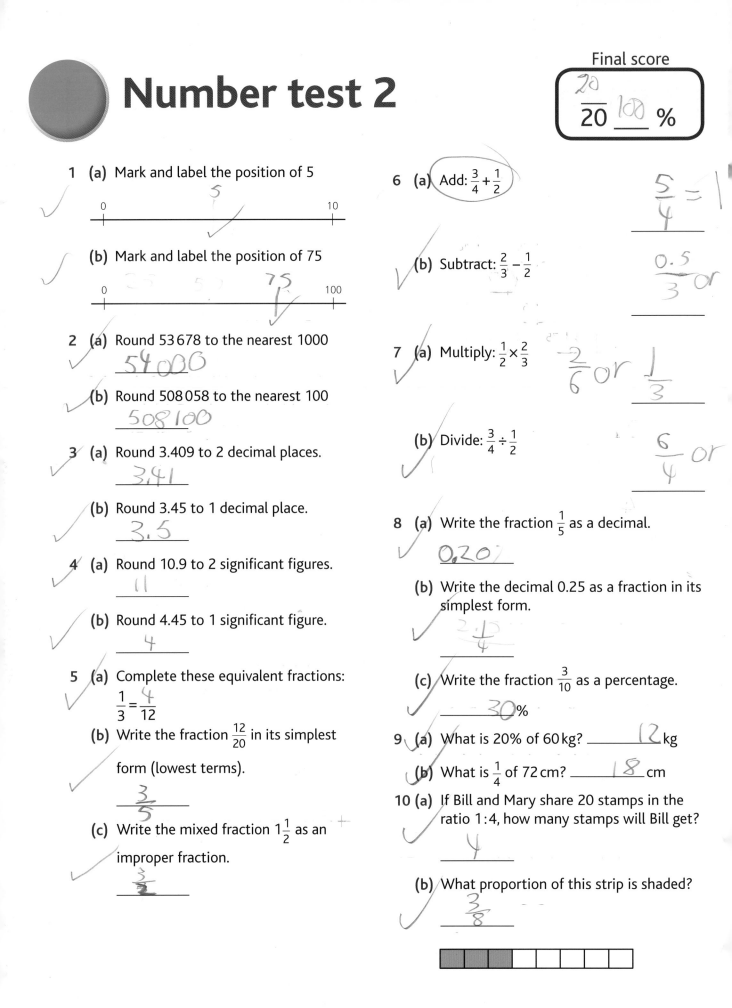

Number test 3

1 (a) On the number grid, shade all the multiples of 4 and circle all the multiples of 11

43	44	45	46	47	48
53	54	55	56	57	58
63	64	65	66	67	68
73	74	75	76	77	78

(b) Write down the number between 30 and 50 which is a multiple of both 6 and 7

__42__

2 (a) What is the highest common factor of 12 and 30?

__6__

(b) Complete the factor rainbow for 48

1 __2__ __3__ __4__ __6__ __8__ __12__ __16__ __24__ 48

3 Write the next two terms in each sequence.

(a) 1, 3, 7, 15, __31__ , __63__

(b) 16, 12, 8, 4, __0__ , __-4__

4 (a) List the prime numbers between 20 and 30 __23, 29__

(b) Write 56 as a product of its prime factors.

__7 × 2 × 2 × 2 = 53__

5 (a) What temperature is 7 degrees lower than 4°C? __-3__ °C

(b) What is the value of ⁻3 + 8? __5__

6 Write down

(a) the square of 8 __64__

(b) the square root of 36 __6__

(c) the cube of 4 __64__

7 (a) Write thirty thousand and five in figures.

__30 005__

(b) Write 201 048 in words.

__two hundred and one__
__thousand and forty eight__

8 (a) Multiply 30 490 by 100 __3 049 000__

(b) Divide 70.4 by 1000 __00,0704__

9 (a) If these decimals are arranged in order of size, which will be in the middle?

__34,52__

32.54   34.52   42.53   23.45   43.25

(b) Write down the number which is exactly half way between 20 and 48

__34__

10 (a) Circle and label the positions 3 and ⁻1

(b) Circle the numbers n such that n > ⁻3

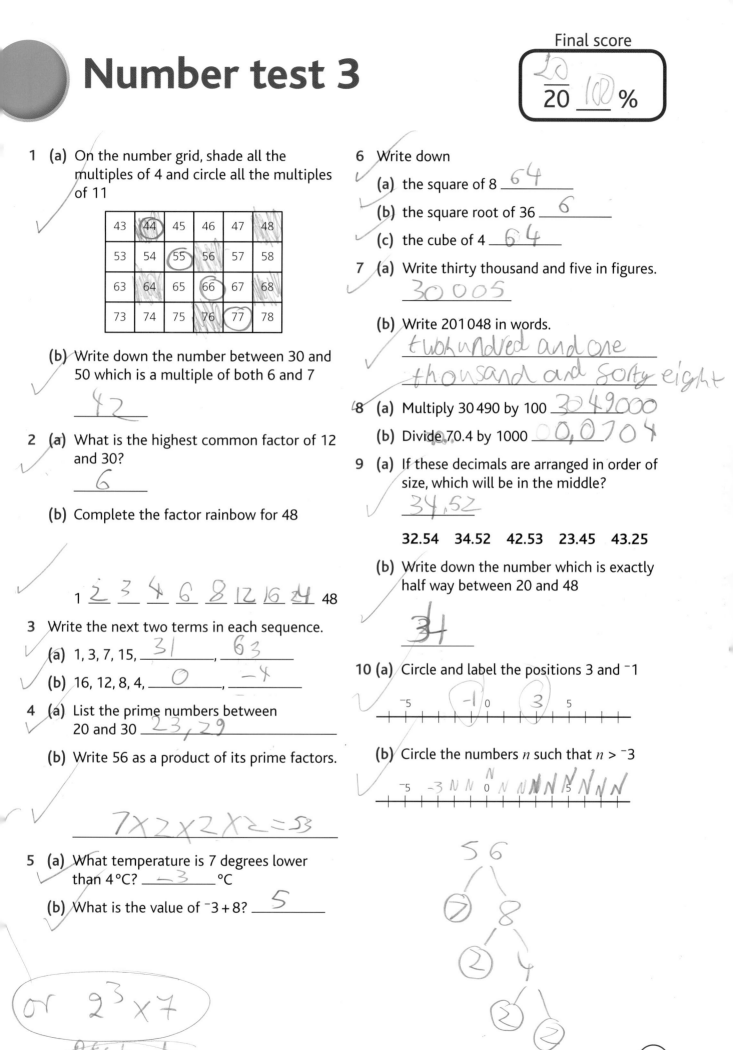

or $2^3 × 7$

? for teacher

9

Number test 4

1 (a) Mark and label the position of 7

```
0                                    10
+------------------------------------+
```

(b) Mark and label the position of 20

```
0                                    100
+------------------------------------+
```

2 (a) Round 230 409 to the nearest 100

_____

(b) Round 949 748 to the nearest 1000

_____

3 (a) Round 45.07 to 1 decimal place.

_____

(b) Round 0.0164 to two decimal places.

_____

4 (a) Round 17.48 to two significant figures.

_____

(b) Round 0.0407 to 1 significant figure.

_____

5 (a) Complete these equivalent fractions:

$$\frac{3}{4} = \frac{}{24}$$

(b) Write the fraction $\frac{16}{40}$ in its simplest form (lowest terms).

_____

(c) Write the improper fraction $\frac{4}{3}$ as a mixed fraction.

_____

6 (a) Add: $\frac{3}{4} + \frac{4}{5}$

_____

(b) Subtract: $\frac{3}{4} - \frac{2}{3}$

_____

7 (a) Multiply: $\frac{3}{4} \times \frac{4}{5}$

_____

(b) Divide: $\frac{3}{4} \div \frac{2}{3}$

_____

8 (a) Write the fraction $\frac{3}{5}$ as a decimal.

_____

(b) Write the decimal 0.45 as a fraction in its simplest form.

_____

(c) Write the fraction $\frac{7}{20}$ as a percentage.

_____ %

9 (a) What is 15% of 40 kg? _____ kg

(b) What is $\frac{3}{5}$ of 45 cm? _____ cm

10 (a) If Tom and Jerry share 40 sweets in the ratio 3:5, how many sweets will Jerry get?

_____

(b) In a garden of area 120 m², there is a lawn of area 80 m². What proportion of the area of the garden is lawn?

_____

Number test 5

1 (a) On the number grid, shade all the multiples of 4 and circle all the multiples of 7

14	15	16	17	18	19
24	25	26	27	28	29
34	35	36	37	38	39
44	45	46	47	48	49

(b) Write down the number between 50 and 60 which is a multiple of both 7 and 4

_____

2 (a) What is the highest common factor of 18 and 54?

_____

(b) Complete the factor rainbow for 40

1 ____ ____ ____ ____ ____ ____ 40

3 Write the next two terms in each sequence.

(a) 1, 5, 9, 13, _____, _____

(b) 1, 4, 10, 22, _____, _____

4 (a) List the prime numbers between 30 and 40 _____

(b) Write 45 as a product of its prime factors.

_____

5 (a) What temperature is 7 degrees lower than 4°C? _____°C

(b) What is the value of ⁻3 + 2? _____

6 Write down

(a) the square of 6 _____

(b) the square root of 64 _____

(c) the cube of 3 _____

7 (a) Write one hundred and two thousand, three hundred and fourteen in figures.

_____

(b) Write 3007 in words.

_____

8 (a) Multiply 4390 by 10 _____

(b) Divide 599 by 1000 _____

9 (a) If these decimals are arranged in order of size, which will be in the middle?

_____

4.82   2.84   4.28   2.48   8.24

(b) Write down the number which is exactly half way between 12 and 42

_____

10 (a) Circle and label the positions 4 and ⁻4

⁻5   0   5
+++++++++++++++++++++++++++++++

(b) Circle the numbers n such that $n < 2$

⁻5   0   5
+++++++++++++++++++++++++++++++

Number test 6

1 (a) Mark and label the position of 2

0 ——————————————————— 10

(b) Mark and label the position of 40

0 ——————————————————— 100

2 (a) Round 244 549 to the nearest 10 000

_____

(b) Round 389 551 to the nearest 1000

_____

3 (a) Round 5.785 to 2 decimal places.

_____

(b) Round 9.04 to 1 decimal place.

_____

4 (a) Round 13.7 to 2 significant figures.

_____

(b) Round 9.86 to 1 significant figure.

_____

5 (a) Complete these equivalent fractions:

$\frac{2}{5} = \frac{}{20}$

(b) Write the fraction $\frac{8}{24}$ in its simplest form (lowest terms).

_____

(c) Write the mixed fraction $1\frac{1}{5}$ as an improper fraction.

_____

6 (a) Add: $\frac{2}{5} + \frac{1}{3}$

_____

(b) Subtract: $\frac{2}{3} - \frac{1}{4}$

_____

7 (a) Multiply: $\frac{1}{3} \times \frac{2}{5}$

_____

(b) Divide: $\frac{2}{3} \div \frac{1}{2}$

_____

8 (a) Write the fraction $\frac{2}{5}$ as a decimal.

_____

(b) Write the decimal 0.15 as a fraction in its simplest form.

_____

(c) Write the fraction $\frac{4}{5}$ as a percentage.

_____ %

9 (a) What is 40% of 20 m? _____ m

(b) What is $\frac{3}{4}$ of 60 litres? _____ litres

10 (a) If Alice and Harry share 50 stickers in the ratio 2:3, how many stickers will Alice get?

_____

(b) What proportion of this strip is shaded?

_____

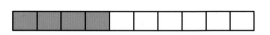

Number test 7

1 (a) On the number grid, shade all the multiples of 3 and circle all the multiples of 8

61	62	63	64	65	66
71	72	73	74	75	76
81	82	83	84	85	86
91	92	93	94	95	96

(b) Write down the number between 30 and 50 which is a multiple of both 3 and 8

_____

2 (a) What is the highest common factor of 14 and 28?

_____

(b) Complete the factor rainbow for 30

1 ____ ____ ____ ____ ____ ____ 30

3 Write the next two terms in each sequence.

(a) 235, 225, 215, 205, _____, _____

(b) 1, 4, 13, 40, _____, _____

4 (a) List the prime numbers between 50 and 60 _____

(b) Write 200 as a product of its prime factors.

_____

5 (a) What temperature is 5 degrees higher than ⁻7°C? _____°C

(b) What is the value of ⁻4 + ⁻3? _____

6 Write down

(a) the square of 9 _____

(b) the square root of 64 _____

(c) the cube of 5 _____

7 (a) Write five hundred and five thousand five hundred and fifty in figures.

_____

(b) Write 44 044 in words.

_____

_____

8 (a) Multiply 14 300 by 10 _____

(b) Divide 48.5 by 10 _____

9 (a) If these decimals are arranged in order of size, which will be in the middle?

_____

567.8   87.65   6.785   687.5   65.78

(b) Write down the number which is exactly half way between 55 and 79

_____

10 (a) Circle and label the positions 2 and ⁻2

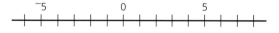

(b) Circle the numbers n such that $n \geq -1$

13

Number test 8

1 (a) Mark and label the position of 3

```
0                              10
+------------------------------+
```

(b) Mark and label the position of 90

```
0                              100
+------------------------------+
```

2 (a) Round 587 595 to the nearest 1000

_____

(b) Round 9539 to the nearest 100

_____

3 (a) Round 5.94 to 1 decimal place.

_____

(b) Round 0.426 to two decimal places.

_____

4 (a) Round 1395 to two significant figures.

_____

(b) Round 0.499 to 1 significant figure.

_____

5 (a) Complete these equivalent fractions:

$$\frac{4}{5} = \frac{}{35}$$

(b) Write the fraction $\frac{12}{50}$ in its simplest form (lowest terms).

_____

(c) Write the improper fraction $\frac{7}{4}$ as a mixed fraction.

_____

6 (a) Add: $\frac{2}{3} + \frac{3}{4}$

_____

(b) Subtract: $\frac{2}{3} - \frac{3}{5}$

_____

7 (a) Multiply: $\frac{2}{3} \times \frac{4}{5}$

_____

(b) Divide: $\frac{2}{3} \div \frac{4}{5}$

_____

8 (a) Write the fraction $\frac{1}{20}$ as a decimal.

_____

(b) Write the decimal 0.35 as a fraction in its simplest form.

_____

(c) Write the fraction $\frac{13}{20}$ as a percentage.

_____ %

9 (a) What is 45% of 80 kg? _____ kg

(b) What is $\frac{3}{4}$ of 1.6 m? _____ m

10 (a) If Ann and Clare share 24 sweets in the ratio 1:3, how many sweets will Clare get?

_____

(b) In a garden of area 100 m², there is a lawn of area 40 m². What proportion of the area of the garden is lawn?

_____

Number test 9

1 (a) On the number grid, shade all the multiples of 5 and circle all the multiples of 4

50	51	52	53	54	55
60	61	62	63	64	65
70	71	72	73	74	75
80	81	82	83	84	85

(b) Write down the smallest number greater than 100 which is a multiple of both 5 and 4

_____

2 (a) What is the highest common factor of 20 and 36?

_____

(b) Complete the factor rainbow for 64

1 _____ _____ _____ _____ _____ 64

3 Write the next two terms in each sequence.

(a) 1, 5, 13, 29, _____ , _____

(b) 8, 4, 2, 1, _____ , _____

4 (a) List the prime numbers between 60 and 70 _____

(b) Write 56 as a product of its prime factors.

_____

5 (a) What temperature is 3 degrees lower than ⁻4°C? _____ °C

(b) What is the value of ⁻5 + 6? _____

6 Write down

(a) the square of 11 _____

(b) the square root of 144 _____

(c) the cube of 6 _____

7 (a) Write seventeen thousand two hundred and six in figures.

_____

(b) Write 21 999 in words.

_____

_____

8 (a) Multiply 58.95 by 10 _____

(b) Divide 98.5 by 100 _____

9 (a) If these decimals are arranged in order of size, which will be in the middle?

_____

75.6   6.75   7.65   65.7   5.76

(b) Write down the number which is exactly half way between 53 and 69

_____

10 (a) Circle and label the positions 2.5 and ⁻3

⁻5   0   5

(b) Circle the numbers n such that $n \leq 1$

⁻5   0   5

Number test 10

1 (a) Mark and label the position of 6

0 ————————————————— 10

(b) Mark and label the position of 60

0 ————————————————— 100

2 (a) Round 28 955 to the nearest 1000

_____

(b) Round 602 783 to the nearest 10 000

_____

3 (a) Round 3.495 to 2 decimal places.

_____

(b) Round 6.06 to 1 decimal place.

_____

4 (a) Round 762 to 2 significant figures.

_____

(b) Round 17.5 to 1 significant figure.

_____

5 (a) Complete these equivalent fractions:

$$\frac{4}{7} = \frac{}{35}$$

(b) Write the fraction $\frac{9}{36}$ in its simplest form

(lowest terms).

_____

(c) Write the mixed fraction $2\frac{1}{3}$ as an

improper fraction.

_____

6 (a) Add: $\frac{2}{7} + \frac{6}{7}$

_____

(b) Subtract: $\frac{2}{5} - \frac{1}{4}$

_____

7 (a) Multiply: $\frac{2}{3} \times \frac{3}{5}$

_____

(b) Divide: $\frac{2}{3} \div \frac{3}{5}$

_____

8 (a) Write the fraction $\frac{1}{8}$ as a decimal.

_____

(b) Write the decimal 0.85 as a fraction in its simplest form.

_____

(c) Write the fraction $\frac{7}{20}$ as a percentage.
_____ %

9 (a) What is 35% of 400 m? _____ m

(b) What is $\frac{4}{5}$ of 60 litres? _____ litres

10 (a) If Kenny and Michelle share 20 sweets in the ratio 4 : 1, how many sweets will Kenny get?

_____

(b) What proportion of this strip is shaded?

_____

Answers

Number test 1

1 (a)

22	23	24	25	26	(27)
32	33	34	35	(36)	37
42	43	44	(45)	46	47
52	53	(54)	55	56	57

(b) 21

2 (a) 4 (b) 2 3 6 9
3 (a) 13 16 (b) 81 243
4 (a) 11 13 17 19 (b) $2^3 \times 3$
5 (a) $^-4\,°C$ (b) 1
6 (a) 16 (b) 3 (c) 8
7 (a) 15 204
  (b) twelve thousand, three hundred and seven
8 (a) 51 750 (b) 4.07
9 (a) 6.75 (b) 18
10 (a) 7 and $^-$3 marked
   (b) all numbers 3 and below marked

Number test 2

1 (a) 5 marked (b) 75 marked
2 (a) 54 000 (b) 508 100
3 (a) 3.41 (b) 3.5
4 (a) 11 (b) 4
5 (a) $\frac{4}{12}$ (b) $\frac{3}{5}$ (c) $\frac{3}{2}$
6 (a) $1\frac{1}{4}$ (b) $\frac{1}{6}$
7 (a) $\frac{1}{3}$ (b) $1\frac{1}{2}$
8 (a) 0.2 (b) $\frac{1}{4}$ (c) 30%
9 (a) 12 kg (b) 18 cm
10 (a) 4 (b) $\frac{3}{8}$ or 37.5%

Number test 3

1 (a)

43	(44)	45	46	47	48
53	54	(55)	56	57	58
63	64	65	(66)	67	68
73	74	75	76	(77)	78

(b) 42

2 (a) 6 (b) 2 3 4 6 8 12 16 24
3 (a) 31 63 (b) 0 $^-4$
4 (a) 23 29 (b) $2^3 \times 7$
5 (a) $^-3\,°C$ (b) 5
6 (a) 64 (b) 6 (c) 64
7 (a) 30 005
  (b) two hundred and one thousand and forty-eight
8 (a) 3 049 000 (b) 0.0704
9 (a) 34.52 (b) 34
10 (a) 3 and $^-$1 marked
   (b) all numbers $^-$2 and above marked

Number test 4

1 (a) 7 marked (b) 20 marked
2 (a) 230 400 (b) 950 000
3 (a) 45.1 (b) 0.02
4 (a) 17 (b) 0.04

5 (a) $\frac{18}{24}$ (b) $\frac{2}{5}$ (c) $1\frac{1}{3}$
6 (a) $1\frac{11}{20}$ (b) $\frac{1}{12}$
7 (a) $\frac{3}{5}$ (b) $1\frac{1}{8}$
8 (a) 0.6 (b) $\frac{9}{20}$ (c) 35%
9 (a) 6 kg (b) 27 cm
10 (a) 25 (b) $\frac{2}{3}$ or 66.$\dot{6}$%

Number test 5

1 (a)

(14)	15	16	17	18	19
24	25	26	27	(28)	29
34	(35)	36	37	38	39
44	45	46	47	48	(49)

(b) 56

2 (a) 18 (b) 2 4 5 8 10 20
3 (a) 17 21 (b) 46 94
4 (a) 31 37 (b) $3^2 \times 5$
5 (a) $^-3\,°C$ (b) $^-1$
6 (a) 36 (b) 8 (c) 27
7 (a) 102 314 (b) three thousand and seven
8 (a) 43 900 (b) 0.599
9 (a) 4.28 (b) 27
10 (a) 4 and $^-$4 marked
   (b) all numbers 1 and below marked

Number test 6

1 (a) 2 marked (b) 40 marked
2 (a) 240 000 (b) 390 000
3 (a) 5.79 (b) 9.0 (0 is essential)
4 (a) 14 (b) 10
5 (a) $\frac{8}{20}$ (b) $\frac{1}{3}$ (c) $\frac{6}{5}$
6 (a) $\frac{11}{15}$ (b) $\frac{5}{12}$
7 (a) $\frac{2}{15}$ (b) $1\frac{1}{3}$
8 (a) 0.4 (b) $\frac{3}{20}$ (c) 80%
9 (a) 8 m (b) 45 litres
10 (a) 20 (b) $\frac{2}{5}$ or 40%

Number test 7

1 (a)

61	62	63	(64)	65	66
71	(72)	73	74	75	76
81	82	83	84	85	86
91	92	93	94	95	(96)

(b) 48

2 (a) 14 (b) 2 3 5 6 10 15
3 (a) 195 185 (b) 121 364
4 (a) 53 59 (b) $2^3 \times 5^2$
5 (a) $^-2\,°C$ (b) $^-7$
6 (a) 81 (b) 8 (c) 125
7 (a) 505 550 (b) forty-four thousand and forty-four
8 (a) 143 000 (b) 4.85
9 (a) 87.65 (b) 67

10 (a) 2 and $^-2$ marked
(b) all numbers $^-1$ and above marked

Number test 8
1 (a) 3 marked (b) 90 marked
2 (a) 588000 (b) 9500
3 (a) 5.9 (b) 0.43
4 (a) 1400 (b) 0.5
5 (a) $\frac{28}{35}$ (b) $\frac{6}{25}$ (c) $1\frac{3}{4}$
6 (a) $1\frac{5}{12}$ (b) $\frac{1}{15}$
7 (a) $\frac{8}{15}$ (b) $\frac{5}{6}$
8 (a) 0.05 (b) $\frac{7}{20}$ (c) 65%
9 (a) 36 kg (b) 1.2 m
10 (a) 18 (b) $\frac{2}{5}$ or 40%

Number test 9
1 (a)

50	51	(52)	53	54	55
(60)	61	62	63	(64)	65
70	71	(72)	73	74	75
(80)	81	82	83	(84)	85

(b) 120

2 (a) 4 (b) 2 4 8 16 32
3 (a) 61 125 (b) $\frac{1}{2}$ $\frac{1}{4}$
4 (a) 61 67 (b) $2^3 \times 7$
5 (a) $^-7$°C (b) 1
6 (a) 121 (b) 12 (c) 216
7 (a) 17206
(b) twenty-one thousand, nine hundred and ninety-nine
8 (a) 589.5 (b) 0.985
9 (a) 7.65 (b) 61
10 (a) 2.5 and $^-3$ marked
(b) all numbers 1 and below marked

Number test 10
1 (a) 6 marked (b) 60 marked
2 (a) 29000 (b) 600000
3 (a) 3.50 (0 essential) (b) 6.1
4 (a) 760 (b) 20
5 (a) $\frac{20}{35}$ (b) $\frac{1}{4}$ (c) $\frac{7}{3}$
6 (a) $1\frac{1}{7}$ (b) $\frac{3}{20}$
7 (a) $\frac{2}{5}$ (b) $1\frac{1}{9}$
8 (a) 0.125 (b) $\frac{17}{20}$ (c) 35%
9 (a) 140 m (b) 48 litres
10 (a) 16 (b) $\frac{1}{3}$ or 33.$\dot{3}$%

Calculations test 1
1 (a) 26 (b) 15 (c) 72 (d) 2
2

×	4	7	6	8
9	36	63	54	72
3	12	21	18	24
5	20	35	30	40
7	28	49	42	56

3 48 − 25 = 23   23 + 25 = 48   25 + 23 = 48
4 30 ÷ 10 = 3   3 × 10 = 30   10 × 3 = 30
5 (a) 43 (b) 23
6 (a) 28 (b) 6501
7 (a) 11 (b) £35.94
8 (a) 16280 (b) 814 (c) 37
9 (a) 537 (b) 105
10 (a) 363 (b) 26.36

Calculations test 2
1 (a) 3045 (b) 48.54
2 5852
3 8.385
4 (a) 142 (b) 3.25
5 (a) 16 remainder 4 (b) 0.43
6 (a) 40 (b) 0.625
7 (a) £40.80 (b) 17
8 (a) 3 hours 30 minutes (b) 3 feet 6 inches
9 (a) odd (b) odd
10 (a) yes (b) yes

Calculations test 3
1 (a) 52 (b) 16 (c) 42 (d) 2
2

×	5	8	7	9
8	40	64	56	72
4	20	32	28	36
6	30	48	42	54
9	45	72	63	81

3 35 − 17 = 18   17 + 18 = 35   18 + 17 = 35
4 56 ÷ 7 = 8   7 × 8 = 56   8 × 7 = 56
5 (a) 27 (b) 30
6 (a) 60 (b) 363
7 (a) 20 (b) £17.94
8 (a) 616 (b) 616 (c) 28
9 (a) 401 (b) 103.2
10 (a) 89 (b) 16.9

Calculations test 4
1 (a) 5085 (b) 17.25
2 7072
3 16.695
4 (a) 19 (b) 2.34
5 (a) 8 remainder 4 (b) 0.44
6 (a) 13 (b) 8.75
7 (a) £56.40 (b) 6
8 (a) 2 hours 36 minutes (b) 2 kg 600 g
9 (a) odd (b) odd
10 (a) no (b) no

Calculations test 5
1 (a) 43 (b) 26 (c) 88 (d) 4
2

×	9	5	4	7
8	72	40	32	56
6	54	30	24	42
7	63	35	28	49
3	27	15	12	21

3 31 − 12 = 19   12 + 19 = 31   19 + 12 = 31

Mathematics Workbook: 10-minute Maths Tests Age 9–11 published by Galore Park

Calculations test 2

Calculators must not be used in this test.

1 (a) Multiply: 435×7

_____

(b) Multiply: 8.09×6

_____

2 Multiply: 308×19

_____

3 Multiply: 6.45×1.3

_____

4 (a) Divide: $994 \div 7$

_____

(b) Divide: $9.75 \div 3$

_____

5 (a) Divide, leaving a remainder: $100 \div 6$

_____

(b) Divide, giving your answer to 2 decimal places: $3 \div 7$

_____

6 (a) Divide, using factors: $560 \div 14$

_____

(b) Divide, giving an exact answer: $5 \div 8$

_____

7 (a) Jo calculated the total cost of her shopping and her calculator display showed the result 40.8

What was the cost of Jo's shopping?

£ _____

(b) Tom calculated the number of bottles he could fill from a large container of juice and his calculator showed the result 17.8

How many bottles would he be able to fill completely?

_____

8 A calculator shows the result 3.5

What would this mean in:

(a) hours and minutes

_____ hours _____ minutes

(b) feet and inches?

_____ feet _____ inches

9 Complete these statements using **even** or **odd**:

(a) An odd number times an _____ number gives an odd number.

(b) An even number minus an odd number gives an _____ number.

10 (a) When Freya added 349 to 47 her calculator showed this result:

| 396 |

Is this correct? _____

(b) When John multiplied 1.5 by 1.2 his calculator showed the result 1.8

Is this correct? _____

Calculations test 1

Calculators must not be used in this test.

1 Write down the

(a) sum of 9 and 17 __16__

(b) difference between 18 and 33
__15__

(c) product of 8 and 9 __72__

(d) remainder when 20 is divided by 3
__1__ 2

2 Complete the multiplication square.

×	4	7	6	8
9	36	63	54	72
3	12	21	18	24
5	20	35	30	40
7	28	49	42	56

3 Write the other three addition and subtraction facts using the same numbers.

48 − 23 = 25

__48__ − __25__ = __23__

__25__ + __23__ = __48__

__23__ + __25__ = __48__

4 Write the other three multiplication and division facts using the same numbers.

30 ÷ 3 = 10

__30__ ÷ __10__ = __3__

__10__ × __3__ = __30__

__3__ × __10__ = __30__

5 Calculate: BIDMAS

(a) 3 + 5 × 8 = 64   43

(b) 4 × 7 − (3 + 2) = __23__

Calculate in your head, with no written working.

6 (a) What is the sum of the first five prime numbers? 1 + 3 + 5 + 7 + 11
__18__ 2

7000 −
+1
6501

(b) Subtract 499 from 7000 6951

7 (a) How many chocolate bars priced at 45p could you buy with a £5 note? 45 × 10 = 4
__10__ 11   45 × 1 = 4

(b) Find the total cost of 6 books priced at £5.99 each.

£ __35.94__

8 Given that 37 × 44 = 1628, complete the following:

(a) 37 × 440 = __16280__

(b) 37 × 22 = __814__

(c) 1628 ÷ 44 = __37__

Set out your working clearly.

9 (a) Add: 450 + 87
450
+87
537   537

(b) Add: 35.7 + 69.3
35 + 70 = 105

10 (a) Subtract: 450 − 87
34⁴15̶0̶
−87
363   363

(b) Subtract: 35.7 − 9.34
23¹5 6⁷¹0
−9 −3 4
26 36   26.36

4   $63 \div 7 = 9$ $7 \times 9 = 63$ $9 \times 7 = 63$
5   (a) 51   (b) 13
6   (a) 25   (b) 1892
7   (a) 10   (b) £39.95
8   (a) 2400   (b) 600   (c) 24
9   (a) 502   (b) 36.75
10   (a) 149   (b) 36.75

Calculations test 6

1   (a) 2529   (b) 399.6
2   10074
3   188
4   (a) 145   (b) 13.1
5   (a) 8 remainder 6   (b) 0.71
6   (a) 29   (b) 5.5
7   (a) £110.50   (b) 6
8   (a) 6 hours 15 minutes   (b) 6 feet 3 inches
9   (a) even   (b) even
10   (a) yes   (b) forgot the point in 3.2

Calculations test 7

1   (a) 170   (b) 114   (c) 84   (d) 2
2

×	7	9	5	12
11	77	99	55	132
8	56	72	40	96
7	49	63	35	84
6	42	54	30	72

3   $40 - 17 = 23$ $23 + 17 = 40$ $17 + 23 = 40$
4   $72 \div 9 = 8$ $8 \times 9 = 72$ $9 \times 8 = 72$
5   (a) 24   (b) 43
6   (a) 68   (b) 138
7   (a) 13   (b) £59.96
8   (a) 434   (b) 896   (c) 15.5
9   (a) 838   (b) 20.35
10   (a) 178   (b) 37.7

Calculations test 8

1   (a) 3531   (b) 117.6
2   29304
3   119.6
4   (a) 36   (b) 1.9
5   (a) 24 remainder 4   (b) 3.14
6   (a) 36   (b) 37.5
7   (a) 76   (b) 5   (c) 4
8   (a) 3 hours 12 minutes   (b) 3 cm 2 mm
9   (a) odd   (b) even
10   (a) place value error   (b) 7

Calculations test 9

1   (a) 210   (b) 54   (c) 132   (d) 3
2

×	8	6	9	7
7	56	42	63	49
11	88	66	99	77
9	72	54	81	63
12	96	72	108	84

3   $88 - 39 = 49$ $88 - 49 = 39$ $39 + 49 = 88$ $49 + 39 = 88$

4   $54 \div 6 = 9$ $54 \div 9 = 6$ $6 \times 9 = 54$ $9 \times 6 = 54$
5   (a) 25   (b) 42
6   (a) 40   (b) 1915
7   (a) 100   (b) £4.95
8   (a) 810   (b) 3240   (c) 22.5
9   (a) 4481   (b) 68.55
10   (a) 244   (b) 36.36

Calculations test 10

1   (a) 2415   (b) 236
2   29088
3   240.24
4   (a) 45   (b) 1.45
5   (a) 14 remainder 2   (b) 3.83
6   (a) 56   (b) 3.7
7   (a) 11 years 9 months   (b) 100 ml
8   (a) 3 hours 3 minutes   (b) 3 kg 50 g
9   (a) even   (b) odd
10   (a) 5.6   (b) forgot to 'carry' one

Problem solving test 1

1   (a)   **Pattern 4**   (b) 5

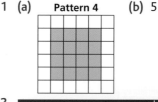

2

Pattern number	1	2	3	4	5
Area (square units)	1	4	9	16	25
Perimeter (units)	4	8	12	16	20

3   (a) 36 square units   (b) 24 units
4   (a) 100 square units   (b) 40 units
5   (a) 8   (b) 400 square units
6   (a) $34 \rightarrow 12 \rightarrow 2$ (b) $77 \rightarrow 49 \rightarrow 36 \rightarrow 18 \rightarrow 8$
   (c) $76 \rightarrow 42 \rightarrow 8$ (d) $99 \rightarrow 81 \rightarrow 8$
7   43
8   46 and 49
9   all give single digit results 5 or zero
10   (a) 17 and 71   (b) because 7 is prime

Problem solving test 2

1   (a) £1.20   (b) £1   10p   10p   or   50p   50p   20p
2   (a) 4%   (b) 3 g
3   (a) 7   (b) 39
4   10
5   50 cm²
6   (a) 200 g   (b) 16
7   (a) 145 hours   (b) 90 hours
8   (a) 40   (b) 6
9   £531
10   Train B; 1 hour 40 minutes

Problem solving test 3

1   (a)   **Pattern 4**   (b) 5

2

Pattern number	1	2	3	4	5
Area (square units)	1	4	9	16	25
Perimeter (units)	4	10	16	22	28

3 (a) 36 square units   (b) 34 units
4 (a) 100 square units   (b) 58 units
5 (a) 20   (b) 144 square units
6 (a) 44 → 24 → 14 → 9   (b) 45 → 25 → 15 → 10 → 5
   (c) 17 → 12 → 7   (d) 99 → 54 → 29 → 19 → 14 → 9
7 41 and 46
8 (a) 0   1   2   3   4   (b) 5 times tens digit is 5 or more
9 they all have digit 5 or 0
10 10   11   12   13   14

Problem solving test 4

1 (a) 47 pence
  (b) Any 4 of: 2 × 20p   2 × 2p   3 × 1p
     20p   10p   3 × 5p   2 × 1p
     4 × 10p   5p   2 × 1p
     3 × 10p   3 × 5p   2p
     20p   2 × 10p   3 × 2p   1p
     20p   5 × 5p   2p
2 (a) 6%   (b) 13 g
3 (a) 5   (b) 9
4 7
5 32 cm
6 (a) 1350 g   (b) 30
7 128 hours
8

	Dogs	No dogs
Boys	4	4
Girls	7	2

9 £332.50

10

Appletown	08:46	10:36	12:26	14:16
Pearville	09:30	11:20	13:10	15:00
Plumford	11:12	13:02	14:52	16:42
Plumford	09:18	11:08	12:58	14:48
Pearville	11:00	12:50	14:40	16:30
Appletown	11:44	13:34	15:24	17:14

Problem solving test 5

1

2

Pattern number	1	2	3	4	5
Number of small triangles	1	4	9	16	25
Perimeter (units)	3	6	9	12	15

3 (a) 36   (b) 18 units
4 (a) 100   (b) 30 units
5 (a) 12   (b) 400
6 (a) 35 → 16 → 14 → 10 → 2   (b) 73 → 20 → 4
   (c) 44 → 16 → 14 → 10 → 2
   (d) 99 → 36 → 18 → 18 … repeats
7 22
8 52 and 53
9 (a) 2   4   6   8   (b) all even numbers
10 (a) 18 and 27   (b) the sum of the digits is 9

Problem solving test 6

1 (a) £3.20
  (b) £2   £1   10p   10p   or   £1   £1   £1   20p
2 (a) 50 g   (b) 1500 g
3 (a) 25   (b) £35
4 3
5 4 m²
6 (a) 24 oz   (b) 3
7 (a) 700 mm   (b) 1000 mm
8 (a) 11   (b) 13
9 £6.25
10 3 hours 55 minutes

Problem solving test 7

1 (a)   **Pattern 4**   (b) 6

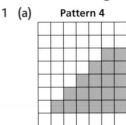

2

Pattern number	1	2	3	4	5
Base length (units)	3	4	5	6	7
Height (units)	2	3	4	5	6
Perimeter (units)	10	14	18	22	26
Area (square units)	5	9	14	20	27

3 (a) 35 square units   (b) 30 units
4 (a) 44 square units   (b) 34 units
5 (a) answers vary, e.g. add base to height then double;
    multiply base by 4 and then subtract 2
  (b) 82 units
6 (a) 12 → 7   (b) 25 → 17 → 22 → 8
   (c) 31 → 6   (d) 99 → 36 → 21 → 5
7 13 and 14
8 20   21   22
9 90
10 58

Problem solving test 8

1 (a) £3.30   (b) £2   £1   20p   10p
2 (a) 80%   (b) 16 g
3 83
4 (a) 11   (b) 11
5 length 40 cm, width 10 cm
6 (a) 270 ml   (b) 340 g
7 2 hours
8 (a) 54   (b) 53%
9 £32
10

Flapford	07:52	09:42	12:32	15:32
Glumton	08:20	10:10	13:00	16:00
Halfly	10:12	12:02	14:52	17:52
Awfly	11:18	13:08	15:58	18:58

Mathematics Workbook: 10-minute Maths Tests Age 9–11 published by Galore Park

Problem solving test 9

1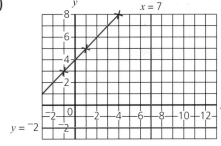

2

Pattern number	1	2	3	4
Green triangles	2	5	9	14
White triangles	1	3	6	10
Total triangles	3	8	15	24

3 (a) 20 (b) 35
4 answers vary, e.g. (pattern number + 1)² − 1 or similar
5 48
6 (a) 17 → 15 → 11 → 3 (b) 82 → 12 → 5
  (c) 68 → 22 → 6   (d) 99 → 27 → 16 → 13 → 7
7 23
8 51 and 54
9 19
10 (a) 38 → 19 → 19 ...   (b) all multiples of 19
       57 → 19 → 19 ...
       76 → 19 → 19 ...

Problem solving test 10

1 (a) £3.10
  (b) £2 50p 3 × 20p or £2 2 × 50p 2 × 5p
     or 3 × £1 5p 5p
2 (a) 87.5 g (b) 12.75 g
3 413
4 15
5 length 3 m, width 1.5 m
6 (a) 1.1718 kg (b) 81
7 3
8 (a) 10 (b) 35%
9 £5.44
10 A

Pre-algebra test 1

1 (a) × 3   − 2 (b) 10 (c) 3
2 (a) multiply by 8 and then divide by 5
  (b) 32 km (c) 35 miles
3 (a) + 4   × 3 (b) 5
4 add 5 then multiply by 2
5 (a) $4a - 3$ (b) 6
6 (a) $g + 4$ (b) $2g$ (c) 7
7 (a) $2a + 2b$ (b) $6a^2b$
8 (a) $a = 9$ (b) $b = 6$ (c) $c = 3$ (d) $d = 36$
9 (a) $e = 2$ (b) $f = 5$
10 (a) $g = 8$ (b) $h = 2$

Pre-algebra test 2

1 (a) 2 (b) 17 (c) 19
2 (a) 13 (b) 24
3 (a) 13 16 + 3 (b) 16 32 × 2
4 (a) 31 × 2 + 1 (b) 41 × 3 − 1
5 7 28 13
6 7 2 6

7 (a)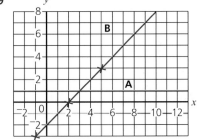

$x = 7$ and $y = {}^-2$ drawn
  (b) $(7, {}^-2)$

8

⁻1	3	(⁻1, 3)
1	5	(1, 5)
4	8	(4, 8)

9 3 points plotted and graph drawn
10 $y = x + 4$

Pre-algebra test 3

1 (a) × 2   + 3 (b) 9 (c) 6
2 (a) multiply by 22 and then divide by 10
  (b) 8.8 lb (c) 50 kg
3 (a) + 3 × 4 (b) 5
4 add 3 then multiply by 3
5 (a) $2g - 5$ (b) 7
6 (a) $s + 5$ (b) $3s$ (c) 10
7 (a) $6a - 3b$ (b) $8a^2b$
8 (a) $a = 8$ (b) $b = 6$ (c) $c = 2$ (d) $d = 16$
9 (a) $e = 7$ (b) $f = 3$
10 (a) $g = 8$ (b) $h = 9$

Pre-algebra test 4

1 (a) 6 (b) 11 (c) 1
2 (a) 29 (b) 39
3 (a) 17 21 + 4 (b) 1 $\frac{1}{2}$ ÷ 2
4 (a) 171 × 4 − 1 (b) 46 × 2 + 2
5 4 16 9
6 4 3 5
7 (a) A $y = 1$ B $x = 5$ (b) (5, 1)
8

5	3	(5, 3)
2	0	(2, 0)
⁻1	⁻3	(⁻1, ⁻3)

9

3 points plotted and graph drawn
10 $y = x - 2$

Pre-algebra test 5

1 (a) × 4 − 1 (b) 11 (c) 4
2 (a) multiply by 5 and then divide by 8
  (b) 10 miles (c) 48 km

3 (a) $+2$ $\times 5$ (b) 4
4 add 3 then multiply by 2
5 (a) $4s-1$ (b) 5
6 (a) $j-3$ (b) $2j$ (c) 6
7 (a) $6a-b$ (b) $3ab^2$
8 (a) $a=12$ (b) $b=3$ (c) $c=2$ (d) $d=50$
9 (a) $e=4$ (b) $f=9$
10 (a) $g=3$ (b) $h=1$

Pre-algebra test 6

1 (a) 1 (b) 8 (c) 9
2 (a) 5 (b) 10
3 (a) 81 243 $\times 3$ (b) 21 26 $+5$
4 (a) 121 $\times 3$ $+1$ (b) 201 $\times 3$ $+3$
5 1 2 5
6 5 2 5
7 (a)

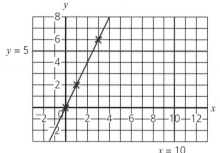

$x=10$ and $y=5$ drawn
  (b) (10, 5)

8

0	0	(0, 0)
1	2	(1, 2)
3	6	(3, 6)

9 3 points plotted and graph drawn
10 $y=2x$

Pre-algebra test 7

1 (a) $\times 4$ $+1$ (b) 9 (c) 6
2 (a) multiply by 3 and then divide by 10
  (b) 1.8 m (c) 8 ft
3 (a) $+4$ $\times 2$ (b) 18
4 add 3 then divide by 2
5 (a) $2(e+3)$ (b) 1.5
6 (a) $t+3$ (b) $3t$ (c) 6
7 (a) $4a+b$ (b) $4a^2b$
8 (a) $a=8$ (b) $b=^-2$ (c) $c=1\frac{1}{2}$ (d) $d=18$
9 (a) $e=4$ (b) $f=3$
10 (a) $g=5$ (b) $h=2$

Pre-algebra test 8

1 (a) 1 (b) 4 (c) 5
2 (a) 3 (b) 14
3 (a) 0 $^-2$ -2 (b) 1 $\frac{1}{3}$ $\div 3$
4 (a) 86 $\times 4$ -2 (b) 50 $\times 2$ -2
5 6 8 2
6 1 4 7
7 (a) A $y=5$ B $x=^-2$ (b) $(^-2, 5)$

8

4	7	(4, 7)
2	5	(2, 5)
$^-2$	1	$(^-2, 1)$

9

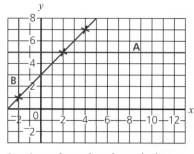

3 points plotted and graph drawn
10 $y=x+3$

Pre-algebra test 9

1 (a) $+4$ $\times 3$ (b) 27 (c) $^-1$
2 (a) multiply by 9 and then divide by 2
  (b) 27 litres (c) 8 gallons
3 (a) $+3$ $\times 4$ (b) 4.5
4 subtract 4 then multiply by 3
5 (a) $5m+2$ (b) 6
6 (a) $d+4$ (b) $2(d+4)$ (c) 4
7 (a) $5a-5b$ (b) a^2b^2
8 (a) $a=9$ (b) $b=^-2$ (c) $c=1\frac{1}{2}$ (d) $d=8$
9 (a) $e=3$ (b) $f=3$
10 (a) $g=^-8$ (b) $h=^-3$

Pre-algebra test 10

1 (a) 2 (b) $^-36$ (c) 6
2 (a) 2 (b) 7
3 (a) 256 1024 $\times 4$ (b) 25 31 $+6$
4 (a) 46 $\times 2$ $+2$ (b) 41 $\times 3$ -1
5 3 $^-2$ $^-1$
6 9 $^-1$ 0.5
7 (a)

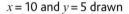

$x=7$ and $y=4$ drawn
  (b) (7, 4)

8

0	0	(0, 0)
6	3	(6, 3)
10	5	(10, 5)

9 3 points plotted and graph drawn
10 $y=\dfrac{x}{2}$

Shape, space and measures test 1

1 (a) 52 mm (b) 0.27 m (c) 4800 g
2 (a) 26 cm (b) 36 cm²
3 (a) 12 units² (b) 16 cm
4 (a) 48 cm³ (b) 80 cm²
5 (a) 18:35 (b) 1.05 pm
  (c) 5 hours 36 minutes (d) 20:25
6 (a) 12 km (b) 1 h 30 min
7 A $^-0.2$ B 1.1 C 1.8 D 5.9

Mathematics Workbook: 10-minute Maths Tests Age 9–11 published by Galore Park

8 **A** isosceles trapezium   **B** rhombus
C isosceles triangle   **D** delta (arrowhead) kite

9

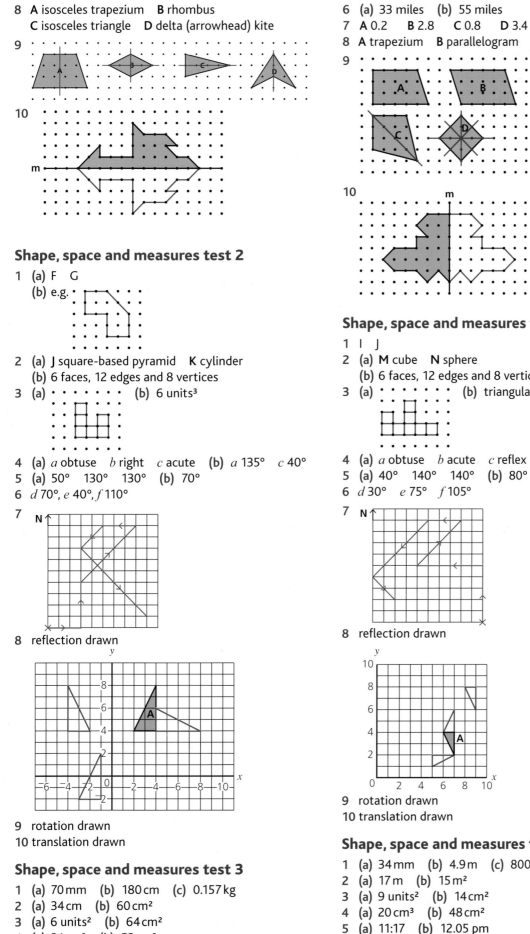

10

Shape, space and measures test 2

1 (a) F   G
   (b) e.g.

2 (a) J square-based pyramid   K cylinder
   (b) 6 faces, 12 edges and 8 vertices

3 (a) ·······   (b) 6 units³

4 (a) *a* obtuse   *b* right   *c* acute   (b) *a* 135°   *c* 40°

5 (a) 50°   130°   130°   (b) 70°

6 *d* 70°, *e* 40°, *f* 110°

7

8 reflection drawn

9 rotation drawn

10 translation drawn

Shape, space and measures test 3

1 (a) 70 mm   (b) 180 cm   (c) 0.157 kg

2 (a) 34 cm   (b) 60 cm²

3 (a) 6 units²   (b) 64 cm²

4 (a) 24 cm³   (b) 52 cm²

5 (a) 20:05   (b) 12.55 pm
   (c) 2 hours 24 minutes   (d) 19:15

6 (a) 33 miles   (b) 55 miles

7 **A** 0.2   **B** 2.8   **C** 0.8   **D** 3.4

8 **A** trapezium   **B** parallelogram   **C** kite   **D** square

9

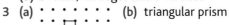

10

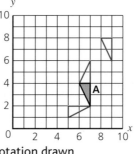

Shape, space and measures test 4

1 I   J

2 (a) M cube   N sphere
   (b) 6 faces, 12 edges and 8 vertices

3 (a) ·······   (b) triangular prism

4 (a) *a* obtuse   *b* acute   *c* reflex   (b) *a* 135°   *c* 225°

5 (a) 40°   140°   140°   (b) 80°

6 *d* 30°   *e* 75°   *f* 105°

7

8 reflection drawn

9 rotation drawn

10 translation drawn

Shape, space and measures test 5

1 (a) 34 mm   (b) 4.9 m   (c) 800 g

2 (a) 17 m   (b) 15 m²

3 (a) 9 units²   (b) 14 cm²

4 (a) 20 cm³   (b) 48 cm²

5 (a) 11:17   (b) 12.05 pm
   (c) 2 hours 48 minutes   (d) 19:23

6 (a) 1.5 km   (b) 40 minutes

7 **A** 0.08   **B** 0.36   **C** ⁻1.2   **D** 3.4

8 **A** isosceles triangle **B** parallelogram
   C trapezium   **D** rectangle

9

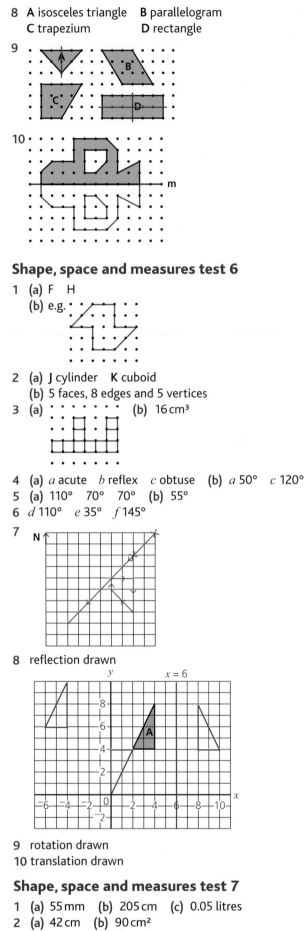

10

---

7 **A** 0.9 **B** 2.7 **C** 0.4 **D** 0.92
8 **A** rhombus **B** isosceles trapezium **C** kite **D** hexagon

9

10

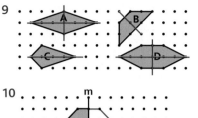

Shape, space and measures test 8

1 J L
2 (a) **M** cuboid   **N** tetrahedron
   (b) 5 faces, 9 edges and 6 vertices
3 (a)   (b) tetrahedron

4 (a) *a* acute   *b* right   *c* obtuse   (b) *a* 40°   *c* 100°
5 (a) 110°   110°   (b) 25°
6 *d* 56°   *e* 62°   *f* 118°
7 **N**

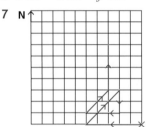

8 reflection drawn

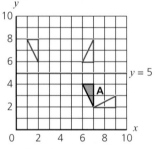

9 rotation drawn
10 translation drawn

Shape, space and measures test 6

1 (a) F   H
   (b) e.g.

2 (a) J cylinder   K cuboid
   (b) 5 faces, 8 edges and 5 vertices
3 (a)   (b) 16 cm³

4 (a) *a* acute   *b* reflex   *c* obtuse   (b) *a* 50°   *c* 120°
5 (a) 110°   70°   70°   (b) 55°
6 *d* 110°   *e* 35°   *f* 145°
7 **N**

8 reflection drawn

9 rotation drawn
10 translation drawn

Shape, space and measures test 7

1 (a) 55 mm   (b) 205 cm   (c) 0.05 litres
2 (a) 42 cm   (b) 90 cm²
3 (a) 12 units²   (b) 12 cm²
4 (a) 16 cm³   (b) 40 cm²
5 (a) 12:45   (b) 1.30 am   (c) 54 minutes   (d) 18:50
6 (a) 5 miles   (b) 0.5 miles

Shape, space and measures test 9

1 (a) 4.4 cm   (b) 1.07 l   (c) 50 g
2 (a) 31 m   (b) 46 m²
3 (a) $4\frac{1}{2}$ units²   (b) 14 cm²
4 (a) 30 cm³   (b) 62 cm²
5 (a) 12:10   (b) 9.59 pm
   (c) 3 hours 54 minutes   (d) 2 hours 55 minutes
6 (a) 3 km   (b) 20 minutes
7 **A** 0.18 **B** 0.52 **C** ⁻1.8 **D** 2.5
8 **A** equilateral triangle   **B** parallelogram
   C kite   **D** regular hexagon

9

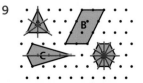

Mathematics Workbook: 10-minute Maths Tests Age 9–11 published by Galore Park

10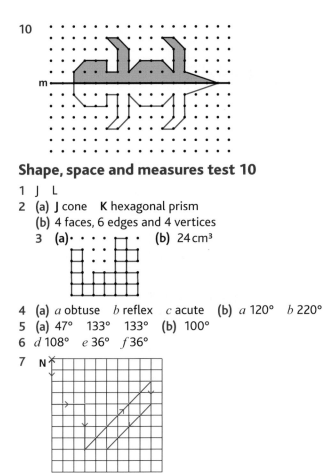

Shape, space and measures test 10

1 J L
2 (a) J cone   K hexagonal prism
  (b) 4 faces, 6 edges and 4 vertices
3 (a) 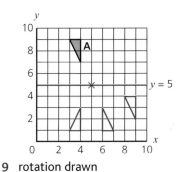   (b) 24 cm³
4 (a) *a* obtuse   *b* reflex   *c* acute   (b) *a* 120°   *b* 220°
5 (a) 47°   133°   133°   (b) 100°
6 *d* 108°   *e* 36°   *f* 36°
7

8 reflection drawn

9 rotation drawn
10 translation drawn

Handling data test 1

1 (a) Charlotte   (b) Daniel
2 (a) 5 months   (b) 40 kg
3

Mark	Tally	Frequency
6	IIII I	6
7	III	3
8	IIII	4
9	II	2

4 (a) 6   (b) 7
5 (a) 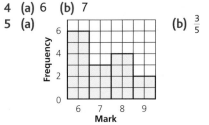   (b) $\frac{3}{5}$

6 A 6   B 2 14   C 9 15   D 1 13
7

	2 14	1 13
Not multiple of 3	2 14	1 13
Multiple of 3	6	9 15
	Even	Not even

8 (a) 7   (b) 4
9 5
10 (a) 27   (b) 5

Handling data test 2

1 (a) 25 km   (b) 57 km
2 (a) $\frac{1}{2}$ (b) 30%
3 (a) 20   (b) 3:2
4 (a) 5 miles   (b) 13 km
5 4 degrees
6 (a) 32   (b) 79
7 2.5
8 (a) D   (b) B
9 (a) D   (b) A
10 (a) even   (b) getting heads with a coin

Handling data test 3

1 (a) Clare   (b) Erica
2 (a) 7 months   (b) 36 kg
3

Mark	Tally	Frequency
4	I	1
5	II	2
6	IIII I	6
7	IIII	5
8	IIII	4
9	II	2

4 (a) 6   (b) 7
5 (a) 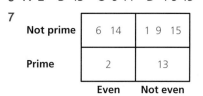   (b) 55%

6 A 2   B 13   C 6 14   D 1 9 15
7

	6 14	1 9 15
Not prime	6 14	1 9 15
Prime	2	13
	Even	Not even

8 (a) 2   (b) Sandra
9 8
10 (a) 13   (b) 16

Handling data test 4

1 (a) 5 cm   (b) 8.5 cm
2 (a) $\frac{1}{4}$ (b) 40%
3 (a) 50   (b) 1:2
4 (a) 2   (b) 7
5 15.5 °C

6 (a) 33 (b) 4
7 104
8 (a) D (b) B
9 (a) D (b) A
10 (a) 12 (b) getting a 6 and a 1

Handling data test 5

1 (a) 2 (b) 40%
2 (a) 2 (b) 3
3

Mark	Tally	Frequency
6	JHI II	7
7	IIII	4
8	JHI I	6
9	III	3

4 (a) 6 (b) 7
5 (a)

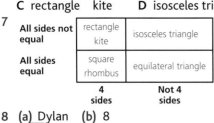

(b) 45%

6 **A** square rhombus **B** equilateral triangle
  C rectangle kite **D** isosceles triangle

7

	4 sides	Not 4 sides
All sides not equal	rectangle kite	isosceles triangle
All sides equal	square rhombus	equilateral triangle

8 (a) Dylan (b) 8
9 11.4
10 (a) 166 (b) 120

Handling data test 6

1 (a) 11 (b) £6.40
2 (a) $\frac{3}{8}$ (b) 7 hours
3 4
4 (a) 6 inches (b) 11 cm
5 (a) 4 degrees (b) 8 °C
6 (a) 20 (b) 55%
7 2
8 (a) B (b) D
9 (a) C (b) A
10 25

Handling data test 7

1 (a) Max (b) Max
2 (a) 1.38 seconds (b) 50 m
3

Score	Tally	Frequency
1	III	3
2	III	3
3	II	2
4	IIII	4
5	JHI	5
6	III	3

4 (a) 5 (b) 4

5 (a)

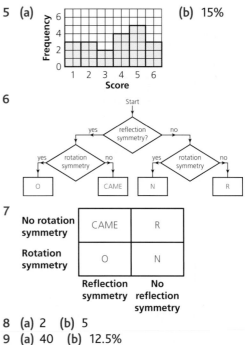

(b) 15%

6

7

	Reflection symmetry	No reflection symmetry
No rotation symmetry	CAME	R
Rotation symmetry	O	N

8 (a) 2 (b) 5
9 (a) 40 (b) 12.5%
10 (a) 7 (b) 14

Handling data test 8

1 (a) 3.3 m (b) 70 cm
2 (a) 15 (b) 50%
3 (a) 6 (b) 3:5
4 (a) 1.2 m (b) 5 feet 3 inches
5 (a) 09:00 and 10:00 (b) 65 cm
6 (a) 25 (b) 3
7 52 8 (a) C (b) A 9 (a) D (b) G
10 arrow indicating position half way between A and B

Handling data test 9

1 (a) 7 months (b) Aaron
2 (a) 41 kg (b) 40 kg
3

Mark	Tally	Frequency
6	IIII	4
7	II	2
8	III	3
9	I	1

4 (a) 6 (b) 7
5 (a)

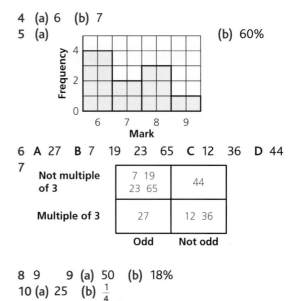

(b) 60%

6 **A** 27 **B** 7 19 23 65 **C** 12 36 **D** 44

7

	Odd	Not odd
Not multiple of 3	7 19 23 65	44
Multiple of 3	27	12 36

8 9 9 (a) 50 (b) 18%
10 (a) 25 (b) $\frac{1}{4}$

Handling data test 10

1 (a) 24 (b) £12.30

2 (a) 20% (b) $\frac{1}{5}$

3 200

4 (a) 2.2 lb (b) 1.4 kg   5 (a) 6 degrees (b) 1°C

6 25%   7 0   8 B   9 (a) E (b) A

10 $\frac{1}{3}$ (3 and 6 are multiples of 3)

Mixed test 1

1 (a) 31 37 (b) $2 \times 3 \times 5$

2 (a) 4.62 (b) 509.03

3 (a) 18 (b) 8.7

4 (a) 0.4 (b) $\frac{3}{4}$ (c) 70%

5

×	7	9	5	6
4	28	36	20	24
8	56	72	40	48
7	49	63	35	42
9	63	81	45	54

6 (a) 10 (b) £9.79 (c) 18:45

7 (a) 6.33 (b) 3.42

8 (a) 2 hours 45 minutes   (b) 2 pounds 12 ounces

9 (a) odd (b) yes

10 (a) 52.8 g (b) 30.8%

Mixed test 2

1 (a) 53→18→11→4   (b) 99→36→15→8

2 41 and 48

3 (a) $a = 13$ (b) $c = 3$

4 1 3 $1\frac{1}{2}$

5

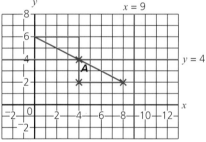

(a) $y = 4$ and $x = 9$ drawn   (b) (9, 4)

6 A $^-$2 B 1.8 C 0.2 D 0.7

7

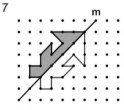

8 (a) points plotted and triangle drawn
(b) triangle rotated

9 (a) 10% (b) $\frac{2}{5}$

10 (a) D (b) A

Mixed test 3

1 (a)

22	23	24	25	26	27
32	33	34	㉟	36	37
㊷	43	44	45	46	47
52	53	54	55	㊶	57

(b) 48

2 (a) 546 (b) 35

3 (a) 11.0 (b) 7.6

4 (a) 180 km (b) 30 kg

5 (a) 15 (b) 41

6 (a) 12 727 (b) 12 727 (c) 1.43

7 16.59

8 (a) 1 day 3 hours   (b) 1 foot $1\frac{1}{2}$ inches

9 (a) odd (b) yes

10 (a) 240 g (b) 18

Mixed test 4

1

Pattern number	1	2	3	4	5
Area (square units)	5	8	11	14	17
Perimeter (units)	12	18	24	30	36

2 (a) 606 (b) ×3 +2

3 (a) $f + 7$ (b) 6

4 3 4 $^-$1

5

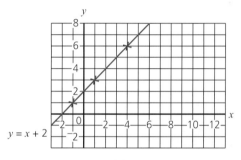

$y = x + 2$

(a) points plotted   (b) line drawn, $y = x + 2$

6 (a) 23 m (b) 28 m²

7 (a) isosceles triangle   (b)

S

8 72° 108° 108°

9 13   10 B

Mixed test 5

1 (a) 41 43 47 (b) $2^3 \times 5$

2 (a) 7.8 (b) 34.05

3 (a) 210 (b) 5.08

4 (a) 0.35 (b) $\frac{4}{5}$ (c) 60%

5

×	6	7	5	9
5	30	35	25	45
9	54	63	45	81
8	48	56	40	72
7	42	49	35	63

6 (a) 15 (b) £23.96 (c) 20:05

7 (a) 8.97   (b) 1.45
8 (a) 1 day 6 hours   (b) 1 foot 3 inches
9 (a) odd   (b) yes
10 (a) 14.7 g   (b) 2.4%

Mixed test 6

1 (a) 23→11→5   (b) 61→25→13→7
2 33 and 39
3 (a) $b = {}^-4$ (b) $d = 45$
4 4   2   4
5

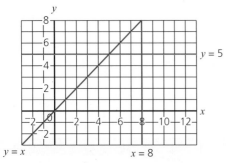

   (a) $x = 8$ and $y = 5$ drawn   (b) (8, 5)   (c) $y = x$ drawn
6 (a) 48 cm³   (b) 80 cm²
7

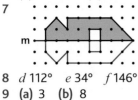

8 d 112° e 34° f 146°
9 (a) 3   (b) 8
10 B

Mixed test 7

1 (a) ⁻3 °C   (b) 3
2 (a) 6.57   (b) 67
3 (a) 19   (b) 9.0
4 (a) $\frac{8}{24}$ (b) $\frac{3}{8}$ (c) $\frac{7}{4}$
5 (a) 43   (b) 23   6 (a) 75   (b) 3511
7 (a) 54.23   (b) 7786
8 (a) £57.40   (b) 4 hours 48 minutes
9 (a) even   (b) odd
10 (a) Y   (b) 5 hours 29 minutes

Mixed test 8

1

Pattern number	1	2	3	4	5
Number of pale squares	5	8	11	14	17
Number of dark squares	1	4	9	16	25

2 (a) 302   (b) 32
3 $2(w + 5)$ or $2w + 10$
4 4   5   ⁻2
5

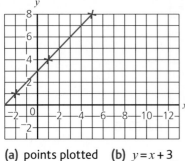

   (a) points plotted   (b) $y = x + 3$

6 (a) 50 mm   (b) 2070 g
7 (a) regular octagon   (b) 8 lines of symmetry drawn
8 37°
9 5
10 C

Mixed test 9

1 (a) 61   67   71   73   79   (b) $2 \times 3 \times 7$
2 (a) 1004   (b) 26.08
3 (a) 44   (b) 10.4
4 (a) 0.8   (b) $\frac{3}{25}$ (c) 55%
5

×	7	11	9	12
8	56	88	72	96
6	42	66	54	72
12	84	132	108	144
11	77	121	99	132

6 (a) 10   (b) £10.89   (c) 19:55
7 (a) 36.9   (b) 41.5
8 (a) 2 hours 3 minutes   (b) 2 kilograms 50 grams
9 (a) odd   (b) yes
10 (a) 13.2 g   (b) 2.4%

Mixed test 10

1 (a) 53 → 20 → 2   (b) 99 → 90 → 9
2 18
3 (a) $a = 12$ (b) $d = 50$
4 3   3   ⁻1
5

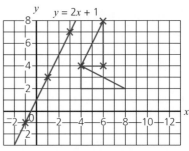

   (a) 3 points plotted   (b) line drawn, $y = 2x + 1$
6 (a) 12 units²   (b) 40 cm²
7

   (a) B   D   F   (b) enlargement drawn
8 (a) points plotted and triangle drawn
   (b) triangle rotated
9 (a) 10%   (b) $\frac{1}{2}$
10 C

Mathematics Workbook: 10-minute Maths Tests Age 9–11 published by Galore Park

Handling data test 1

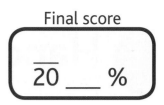
Questions 1 and 2 refer to this table of data.

Name	Age (y:m)	Height (cm)	Mass (kg)
Amelia	11:01	140	41
Billy	10:11	138	38
Charlotte	11:02	135	39
Daniel	10:09	137	42

1 (a) Who was born first? _____

(b) When the children are arranged in order of height, who will stand next to Charlotte? _____

2 (a) What is the range of ages? _____ months

(b) What is the mean mass? _____ kg

Questions 3 to 5 refer to this list of the marks 15 pupils gained in a test.

8 7 6 6 8 9 6 6 8 7 6 9 7 8 6

3 Complete the tally and frequency table.

Mark	Tally	Frequency
6		
7		
8		
9		

4 (a) What is the modal mark? _____

(b) What is the median mark? _____

5 (a) Complete the frequency diagram.

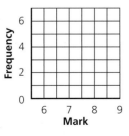

(b) What fraction of the pupils got 7 or more marks? _____

Questions 6 and 7 refer to these numbers.

1 2 6 9 13 14 15

6 Use the flowchart to sort the numbers into the correct boxes A, B, C and D.

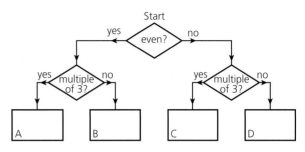

7 Write the numbers in the Carroll diagram below.

Questions 8 and 9 refer to the pictogram below showing the results of a survey into air travel.

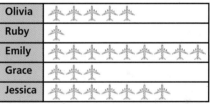

One symbol represents one return flight.

8 (a) How many return flights has Jessica made? _____

(b) How many more return flights than Olivia has Emily made? _____

9 What was the mean number of return flights made by a person? _____

10 The Venn diagram shows the numbers of pupils in Year 6 who have played in the football team.

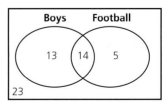

(a) How many boys are in Year 6? _____

(b) How many girls have played in the team? _____

Handling data test 2

1 The bar chart shows the lengths of three rivers.

Length (km)

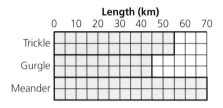

(a) How much longer is the Meander than the Gurgle? _____ km

(b) What is the average (mean) length, to the nearest km, of the rivers? _____ km

Questions 2 to 3 refer to the diagram below showing the colours of 100 flags.

2 (a) What fraction of the flags is red? _____

(b) What percentage of the flags is blue? _____ %

3 (a) How many more red flags are there than blue flags? _____

(b) What is the ratio, in its simplest form, of blue flags to yellow flags? _____ : _____

4 The graph below converts kilometres to miles.

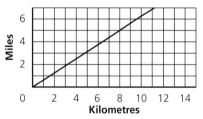

(a) How many miles are equivalent to 8 kilometres? _____ miles

(b) How many kilometres (to the nearest kilometre) are equivalent to 8 miles? _____ km

5 The line graph shows the temperature recorded every half hour.

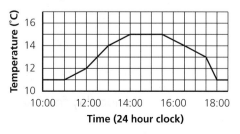

How many degrees warmer was it at 14:00 than at 11:00? _____ degrees

Questions 6 and 7 refer to the bar line graph which shows the results of spinning a fair square spinner numbered 1 to 4

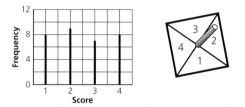

6 (a) How many times was the spinner spun? _____

(b) What was the total of all the scores? _____

7 What was the mean score, to 1 decimal place? _____

Questions 8 to 10 refer to these items.

8 On the scale below, which letter best represents the likelihood of getting

(a) 'heads' with the coin _____

(b) 6 when the die is rolled? _____

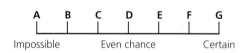

9 On the scale above, which letter best represents the likelihood of getting

(a) an even number when the die is rolled _____

(b) a total of less than 2 if the die is rolled twice? _____

10 (a) If the coin is tossed twice and it falls heads both times, what is the chance that it will fall tails next time? _____

(b) Which is more likely, getting heads with the coin or getting a square number with the die? _____

Handling data test 3

Questions 1 and 2 refer to this table of data.

Name	Age (y:m)	Height (m)	Mass (kg)
Alan	10:11	1.32	35
Becky	11:02	1.41	37
Clare	11:03	1.30	33
Dan	10:10	1.37	41
Erica	10:08	1.35	34

1 (a) Who was born first? _____

(b) When the children are arranged in order of height with the tallest first, who will be in the middle? _____

2 (a) What is the range of ages? _____ months

(b) What is the mean mass? _____ kg

Questions 3 to 5 refer to this list of the marks 20 pupils gained in a test.

6 8 9 6 5 7 6 6 8 7

6 9 7 8 4 5 8 7 7 6

3 Complete the tally and frequency table.

Mark	Tally	Frequency
4		
5		
6		
7		
8		
9		

4 (a) What is the modal mark? _____

(b) What is the median mark? _____

5 (a) Complete the frequency diagram.

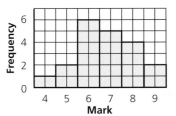

(b) What percentage of the pupils got 7 or more marks? _____ %

Questions 6 and 7 refer to these numbers.

1 2 6 9 13 14 15

6 Use the flowchart to sort the numbers into the correct boxes A, B, C and D.

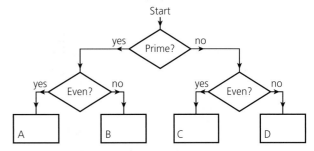

7 Write the numbers in the Carroll diagram below.

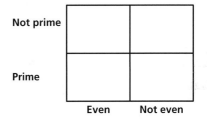

Questions 8 and 9 refer to the pictogram below which shows the results of a search for crabs.

Philip	🦀 🦀 🦀 🦀 🦀
Quentin	🦀 🦀 🦀 🦀
Rachel	🦀 🦀
Sandra	🦀 🦀 🦀 🦀 🦀 🦀
Trevor	🦀 🦀 🦀 🦀

Philip found two more crabs than Quentin.

8 (a) How many crabs are represented by each symbol? _____

(b) Who found 4 more crabs than Trevor? _____

9 What was the mean number of crabs found by one person? _____

10 The Venn diagram shows the numbers of pupils who play hockey and football.

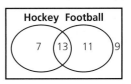

(a) How many pupils play both sports? _____

(b) How many do not play football? _____

Handling data test 4

1 The bar chart shows the lengths of 4 pencils.

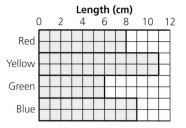

Length (cm)

(a) How much longer is the yellow pencil than the green pencil? _____ cm

(b) What is the average (mean) length of a pencil? _____ cm

Questions 2 to 3 refer to the diagram below showing the colours of 200 balls.

2 (a) What fraction of the balls is red? _____

(b) What percentage of the balls is green? _____ %

3 (a) How many more green balls are there than blue balls? _____

(b) What is the ratio, in its simplest form, of yellow balls to green balls? _____ : _____

4 The graph below converts booboos to blips.

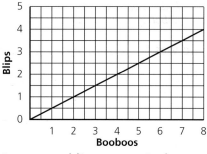

(a) How many blips are equivalent to 4 booboos? _____

(b) How many booboos are equivalent to 3.5 blips? _____

5 The line graph shows the temperature recorded every 30 minutes.

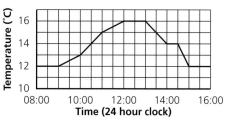

What was the temperature at 11:30? _____ °C

Questions 6 and 7 refer to the bar line graph which shows the results of spinning a fair pentagonal spinner numbered 1 to 5

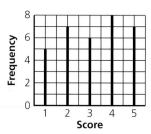

6 (a) How many times was the spinner spun? _____

(b) What was the modal score? _____

7 What was the total of all the scores? _____

Questions 8 to 10 refer to these items.

8 On the scale below, which letter best represents the likelihood of getting

(a) 'heads' with the coin _____

(b) 3 when one die is rolled? _____

A   B   C   D   E   F   G

Impossible   Even chance   Certain

9 On the scale above, which letter best represents the likelihood of getting

(a) a prime number when one die is rolled _____

(b) a total of more than 12 if both dice are rolled? _____

10 (a) If, at the same time, the coin is tossed and one die is rolled, what is the total number of possible outcomes? _____

(b) If, at the same time, both dice are rolled, which is more likely, getting two sixes or getting a six and a one? _____

Handling data test 5

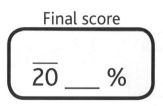
Questions 1 and 2 refer to this table of data.

Name	Favourite sport	Number of pets	Eye colour
Thomas	football	2	blue
Harry	hockey	3	blue
Chloe	football	2	brown
Sophie	football	7	blue
Joshua	hockey	1	brown

1 (a) How many friends with blue eyes prefer football to hockey? _____

 (b) What percentage of the friends have brown eyes? _____ %

2 (a) What number of pets is the mode? _____

 (b) What is the mean number of pets? _____

Questions 3 to 5 refer to this list of the marks 20 pupils gained in a test.

9 6 6 8 8 7 6 6 8 7

6 9 7 8 6 9 7 8 6 8

3 Complete the tally and frequency table.

Mark	Tally	Frequency
6		
7		
8		
9		

4 (a) What is the modal mark? _____

 (b) What is the median mark? _____

5 (a) Complete the frequency diagram.

 (b) What percentage of the pupils got 8 or more marks? _____ %

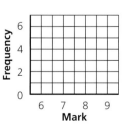

Questions 6 and 7 refer to this list of shapes.

Square   Rectangle   Isosceles triangle

Rhombus   Kite   Equilateral triangle

6 Use the flowchart to sort the shapes into the correct boxes A, B, C and D.

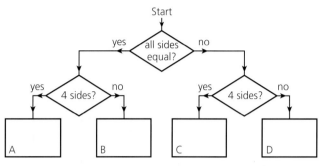

7 Write the names in the Carroll diagram below.

	4 sides	Not 4 sides
All sides not equal		
All sides equal		

Questions 8 and 9 refer to the pictogram showing the goals scored one season.

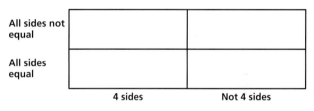

One symbol ⚽ represents two goals.

8 (a) Who was the top goal scorer? _____

 (b) How many more goals did Charlie score than Ethan? _____

9 What was the mean number of goals scored by a player? _____

10 The Venn diagram shows the numbers of pupils who are in the choir and orchestra.

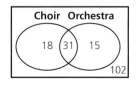

 (a) How many pupils are in the school? _____

 (b) How many pupils are not in the orchestra? _____

Handling data test 6

1 The bar chart records the coins in Mia's piggybank.

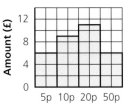

(a) How many 20p coins does Mia have?
_____

(b) What is the total value of Mia's coins?
£ _____

Questions 2 and 3 refer to the diagram showing a typical 24 hours for Eva.

Eva sleeps for 9 hours.

Hours spent

Key
- Sleeping
- Working
- Eating
- Sports
- Resting

2 (a) What fraction of the day does Eva spend sleeping? _____

(b) For 8 hours Eva is not working or sleeping. How long does Eva work? _____ hours

3 The 8 hours Eva is not working or sleeping is shared between eating, sports and resting in the ratio 1 : 2 : 1

How many hours does she do sports? _____

4 The graph below converts centimetres to inches.

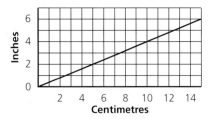

Centimetres

(a) How many inches are equivalent to 15 centimetres? _____ inches

(b) How many centimetres are equivalent to $4\frac{1}{2}$ inches? _____ cm

5 The table records the temperature in degrees Celsius at midday one week.

Day	Sun	Mon	Tue	Wed	Thu	Fri	Sat
Temperature	6	7	9	10	9	7	8

(a) What was the range of midday temperatures? _____ degrees

(b) What was the mean midday temperature that week? _____ °C

Questions 6 and 7 refer to the bar line graph which shows the results when Evie tossed a coin.

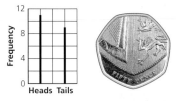

Heads Tails

6 (a) How many times was the coin tossed?
_____

(b) What percentage of the tosses were 'heads'? _____ %

7 If Evie scores 1 point for every 'head' and loses a point for every 'tail', what is her total score? _____

Questions 8 to 10 refer to these spinners.

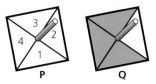

P   Q

8 On the scale below, which letter best represents the likelihood of getting

(a) 4 with spinner P _____

(b) 'red' with spinner Q? _____

A   B   C   D   E

Impossible   Even chance   Certain

9 On the scale above, which letter best represents the likelihood of getting

(a) a square number with spinner P _____

(b) a multiple of 5 with spinner P? _____

10 If spinner Q is spun 100 times, how many times would you expect it to fall on 'white'?
_____

Handling data test 7

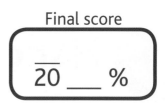
Final score

$\overline{20 \quad}$ %

Questions 1 and 2 refer to this table of data.

Name	Age (y:m)	100 m time (seconds)	Cricket ball throw (m)
Megan	10:10	15.33	53
Jacob	11:00	15.51	47
Katie	10:09	15.96	42
Max	11:02	16.24	58
Holly	10:11	16.71	50

1 (a) Who is oldest? _____

(b) Who made the best cricket ball throw? _____

2 (a) What is the range of 100 m times? _____ seconds

(b) What is the mean cricket ball throw? _____ m

Questions 3 to 5 refer to this list of 20 scores when Tyler rolled a die.

6 3 5 5 2 1 4 6 4 5
6 1 2 4 4 5 3 2 1 5

3 Complete the tally and frequency table.

Score	Tally	Frequency
1		
2		
3		
4		
5		
6		

4 (a) Which score is the mode? _____

(b) What is the median score? _____

5 (a) Complete the frequency diagram.

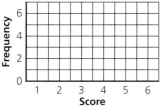

(b) What percentage of the scores were 6? _____ %

Questions 6 and 7 refer to the symmetry of the letters in Cameron's name.

C   A   M   E   R   O   N

6 Use the flowchart to sort the letters into the correct boxes.

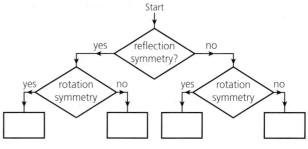

7 Write the letters in the Carroll diagram below.

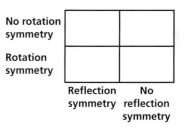

	Reflection symmetry	No reflection symmetry
No rotation symmetry		
Rotation symmetry		

Questions 8 and 9 refer to the pictogram which shows the numbers of cats living in a village.

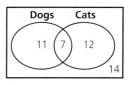

Tabby	
Ginger	
Black	
Others	

There are 2 more black cats than tabby cats.

8 (a) How many cats are represented by each symbol? _____

(b) How many ginger cats are there? _____

9 (a) What is the total number of cats? _____

(b) What percentage of the cats is ginger? _____ %

10 The Venn diagram shows the dog and cat ownership of pupils in Year 6

(a) How many pupils own both dogs and cats? _____

Dogs   Cats

11   7   12

14

(b) How many do not own dogs or cats? _____

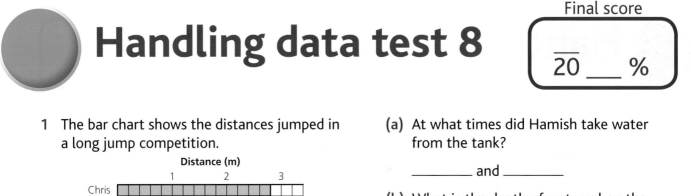

Handling data test 8

1 The bar chart shows the distances jumped in a long jump competition.

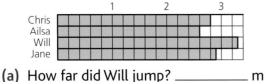

Distance (m)

(a) How far did Will jump? _____ m

(b) What is the range? _____ cm

Questions 2 to 3 refer to the diagram below showing the types of 48 sweets.

Mints   Toffees   Chocolates

2 (a) How many toffees are there? _____

  (b) What percentage of the sweets are chocolates? _____ %

3 (a) How many more toffees are there than mints? _____

  (b) What is the ratio of mints to toffees? _____ : _____

4 The graph below converts feet to metres.

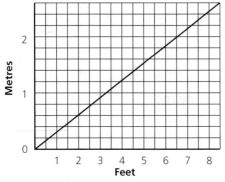

Feet

(a) How many metres are equivalent to 4 feet? _____ m

(b) How many feet are equivalent to 1.6 metres? _____ feet _____ inches

5 The line graph shows the depth of water in Hamish's water tank.
The tank was full at 08:00

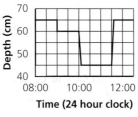

Time (24 hour clock)

(a) At what times did Hamish take water from the tank?

_____ and _____

(b) What is the depth of water when the tank is full?

_____ cm

Questions 6 to 8 refer to the bar line graph which shows the results of spinning the spinner pictured below.

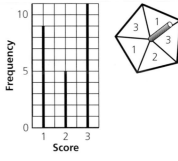

Score

6 (a) How many times was the spinner spun?

_____

  (b) What was the modal score? _____

7 What was the sum of all the scores? _____

8 With the spinner above, which letter on the scale below best represents the likelihood of getting

  (a) 1 _____

  (b) 4? _____   Impossible   Certain

Questions 9 and 10 refer to these items.

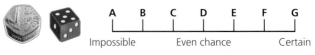

9 On the scale above, which letter best represents the likelihood of getting

  (a) a prime number with the die _____

  (b) a total of 3 or more if the die is rolled 3 times? _____

10 If, at the same time, the coin is tossed and the die is rolled, mark with an arrow on the scale the likelihood of getting heads and 6.

Handling data test 9

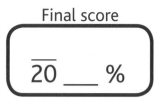
Questions 1 and 2 refer to this table of data.

Name	Age (y:m)	Height (m)	Mass (kg)
Logan	10:09	1.43	41
Rhys	11:01	1.51	42
Ella	10:11	1.38	42
Aaron	11:03	1.40	40
Daisy	10:08	1.39	35

1 (a) What is the range of ages? _____ months

   (b) If the five pupils stood in order of increasing height, who would be in the middle? _____

2 (a) What is the median mass? _____ kg

   (b) What is the mean mass? _____ kg

Questions 3 to 5 refer to this list of Matilda's marks for 10 tests.

6 6 7 8 8 6 9 7 8 6

3 Complete the tally and frequency table.

Mark	Tally	Frequency
6		
7		
8		
9		

4 (a) What is the modal mark? _____

   (b) What is the median mark? _____

5 (a) Complete the frequency diagram.

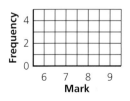

   (b) What percentage of Matilda's marks were 7 or more? _____ %

Questions 6 and 7 refer to these numbers.

7 12 19 23 27 36 44 65

6 Use the flowchart to sort the shapes into the correct boxes A, B, C and D.

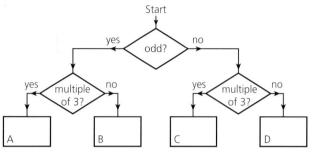

7 Write the numbers in the Carroll diagram below.

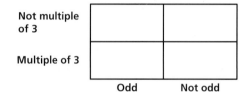

Questions 8 and 9 refer to the pictogram showing the dogs seen in a park one day.

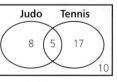

One symbol 🐕 represents two dogs.

8 How many more Collies were seen than Alsatians? _____

9 (a) What was the total number of dogs seen in the park that day? _____

   (b) What percentage of the dogs seen were Spaniels? _____ %

10 The Venn diagram shows the numbers of children who are in the judo and tennis clubs.

   (a) How many children are in just one of these clubs? _____

   (b) What fraction of the children are not in either club? _____

Handling data test 10

1   The bar chart records the coins put in a charity box.

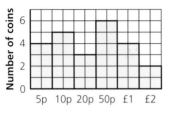

   (a) What is the total number of coins?
   _____

   (b) What is the total value of the coins?
   £ _____

Questions 2 and 3 refer to the pie chart which shows the sales in a shop selling Scottish dress. Kilts make up 40% of the items sold and a third of the remaining items are jackets.
The number of belts and extras sold were the same and together make up 20% of the total number of items sold.

Number of items sold

Key
▨ Kilts
☐ Jackets
▨ Sporrans
▨ Belts
▨ Extras

2   (a) What percentage of the items were jackets? _____%

   (b) What fraction of the items sold was sporrans? _____

3   If 80 kilts were sold, how many items were sold altogether? _____

4   The graph below converts pounds to kilograms.

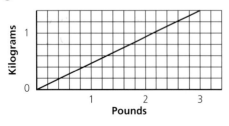

Pounds

   (a) How many pounds are equivalent to 1 kg?
   _____ lb

   (b) How many kilograms are equivalent to 3 pounds? _____ kg

5   The table records the temperature in degrees Celsius at midday one week.

Day	Sun	Mon	Tue	Wed	Thu	Fri	Sat
Temperature (°C)	⁻2	⁻1	0	1	3	2	4

   (a) What was the range of midday temperatures? _____ degrees

   (b) What was the mean midday temperature that week? _____ °C

Questions 6 to 8 refer to the bar line graph showing how a model car landed when Ethan dropped it.

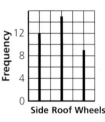 

6   In what percentage of the drops did the car land on its wheels? _____ %

7   If Ethan scores 2 points for every 'wheels', 1 for every 'side' and loses 2 points for every 'roof', what is his total score? _____

8   On the scale below, which letter best represents the likelihood that the car will land on its wheels. _____

9   On the scale above, which letter best represents the likelihood of you:

   (a) growing at least 1 mm in the next year
   _____

   (b) running 100 m in less than 9 seconds?
   _____

10 Describe, as best you can, the likelihood of scoring a multiple of 3 when an ordinary die is rolled. _____

Mixed test 1

1 (a) List the prime numbers between
        30 and 40 _____

   (b) Write 30 as a product of its prime factors.

        _____

2 (a) Multiply 0.462 by 10 _____

   (b) Divide 50 903 by 100 _____

3 (a) Round 17.95 to two significant figures.

        _____

   (b) Round 8.749 to 1 decimal place.

        _____

4 (a) Write the fraction $\frac{2}{5}$ as a decimal.

        _____

   (b) Write the decimal 0.75 as a fraction in its
        simplest form.

        _____

   (c) Write the fraction $\frac{7}{10}$ as a percentage.
        _____ %

5 Complete the multiplication square.

×	7	9	5	6
4				
8				
7				
9				

6 (a) How many magazines priced at £1.95
        could you buy with a £20 note?

        _____

   (b) Find the total cost of 11 chocolate bars
        priced at 89p each.

        £ _____

   (c) At what time did a $2\frac{1}{2}$ hour film start if it

        ended at 21:15? _____ : _____

7 (a) Subtract: 10.07 – 3.74

        _____

   (b) Divide: 23.94 ÷ 7

        _____

8 A calculator shows the result: 2.75

   What would this mean in:

   (a) hours and minutes

        _____ hours _____ minutes

   (b) pounds and ounces?

        _____ pounds _____ ounces

9 (a) Complete this statement using **even**
        or **odd**:

        An odd number plus an _____
        number gives an even number.

   (b) When Justyna multiplied 40.5 by 8 her
        calculator showed the result: 324
        Is this correct?

        _____

10 The nutrition information on a 25 g packet of
   crisps is:

	per 25 g
Fat	7.7 g
Carbohydrate	13.2 g
Fibre	1.1 g
Protein	1.6 g
Salt	0.3 g

   (a) What would be the mass of carbohydrate
        in 100 grams of these crisps?

        _____ g

   (b) What percentage of a crisp is fat?

        _____ %

Mixed test 2

Questions 1 and 2 concern sequences generated from 2-digit starting numbers by applying a rule over and over again.

The rule is: Multiply the tens digit by 3 and then add the units digit. Stop when you get a single digit. For example:

$45 \rightarrow 17 \rightarrow 10 \rightarrow \underline{3}$ (3 steps)

1 Write down the sequences for the following starting numbers:

   (a) 53 _____

   (b) 99 _____

2 Find two starting numbers between 40 and 50 which result in the single digit 6

   _____ and _____

3 Solve the equations:

   (a) $a - 5 = 8$ $a =$ _____

   (b) $3c = 9$ $c =$ _____

4 Complete the table of input and output values for this function machine.

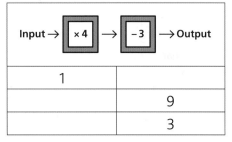

1	
	9
	3

5 (a) On the grid, draw the lines $y = 4$ and $x = 9$

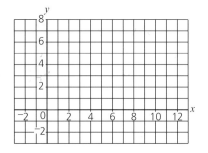

   (b) Write the co-ordinates of the point where the lines $y = 4$ and $x = 9$ cross. (____ , ____)

6 What are the readings on the scales?

   A _____   B _____

   C _____   D _____

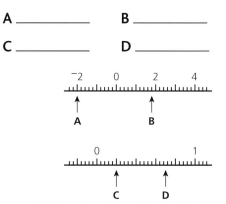

7 Complete the shape which is symmetrical about the line **m**.

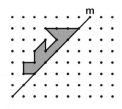

8 (a) On the grid for question 5, plot the points (4, 2), (8, 2) and (4, 4). Join the points to form triangle **A**.

   (b) Rotate triangle **A** through 180° about the point (4, 4).

9 The diagram shows the proportions of exam grades awarded to 40 children.

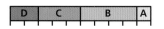

   (a) What percentage of the children achieved A grades? _____ %

   (b) What fraction, in its simplest form, achieved B grades?

   _____

10 On the scale below, which letter best represents the likelihood of getting

A   B   C   D   E   F   G
Impossible   Even chance   Certain

   (a) 'heads' when the coin is tossed

   _____

   (b) 4 when the spinner is spun? _____

Mixed test 3

1 (a) On the number grid, shade all the multiples of 4 and circle all the multiples of 7

22	23	24	25	26	27
32	33	34	35	36	37
42	43	44	45	46	47
52	53	54	55	56	57

(b) Write down the number between 30 and 50 which is a multiple of both 6 and 8

_____

2 (a) If these numbers are arranged in order of size, which will be in the middle?

_____

456   654   546   564   465

(b) Write down the number which is exactly half way between 22 and 48 _____

3 (a) Round 10.95 to 1 decimal place.

_____

(b) Round 7.55 to 2 significant figures.

_____

4 (a) What is 75% of 240 km?
_____ km

(b) What is $\frac{2}{5}$ of 75 kg?
_____ kg

5 Calculate:

(a) $(7 - 4) \times 5$

_____

(b) $6 \times 9 - (5 + 8)$

_____

6 Given that $8.9 \times 14.3 = 127.27$, complete the following:

(a) $89 \times 143 =$ _____

(b) $890 \times 14.3 =$ _____

(c) $127.27 \div 89 =$ _____

7 Multiply: 3.95×4.2

_____

8 A calculator shows the result: [1.125]

What would this mean in:

(a) days and hours

_____ day _____ hours

(b) feet and inches?

_____ foot _____ inches

9 (a) Complete this statement using **even** or **odd**:

An even number minus an _____ number gives an odd number.

(b) When Freddie divided 124 by 8 his calculator showed the result: [15.5]

Is this correct? _____

10 Mary has the recipe below and wants to bake 24 scones.

To make half a dozen scones			
Flour	240 g	Milk	100 ml
Butter	50 g	Baking powder	3 tsp
Caster sugar	35 g	Salt	$\frac{1}{2}$ tsp
Sultanas	60 g	Eggs	1

(a) What mass of sultanas will Mary need?

_____ g

Mary finds that she has only 720 grams of flour, but plenty of the other ingredients.

(b) How many scones could she bake?

_____

Mixed test 4

Questions 1 and 2 concern this pattern sequence.

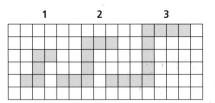

1 Complete this table of data.

Pattern number	1	2	3	4	5
Area (square units)	5	8			
Perimeter (units)	12	18			

2 (a) Alice suggests that you can use the flowchart below to find the perimeter.

Pattern number → ×6 → +6 → Perimeter

Use Alice's idea to calculate the perimeter of pattern 100 _____ units

(b) Complete this flowchart to find the area.

Pattern number → × ☐ → + ☐ → Area

3 Flora has f pens. George has 7 more pens than Flora.

(a) Write an expression, in terms of f, for the number of pens that George has.

_____

(b) Harry has 18 pens and the three friends between them have a total of 37 pens. How many pens does Flora have?

_____

4 Complete the table of input and output values for this machine.

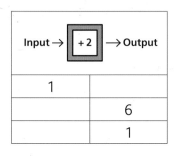

Input → +2 → Output

1	
	6
	1

5 (a) On the grid, plot the input (x) and output (y) values for the machine in question 4

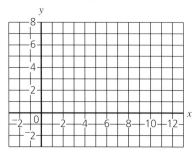

(b) Draw the graph of the function and complete the equation of the line.

$y = $ _____

6 For this rectangle calculate:

8 m
3.5 m

(a) the perimeter _____ m

(b) the area. _____ m²

7 (a) Name the plane shape **S**. _____

S

(b) Mark all lines of symmetry on shape **S**.

8 One angle of an isosceles trapezium is 72°. What sizes are the other three angles?

_____ ° _____ ° _____ °

9 The Venn diagram below shows the numbers of pupils who have dogs and cats.

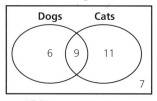
Dogs   Cats
6 9 11
7

How many do not have cats? _____

10 Two coins are tossed at the same time. Which letter on the scale best represents the likelihood of getting two heads?

_____

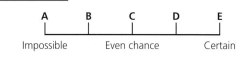
A   B   C   D   E
Impossible   Even chance   Certain

Mixed test 5

1 (a) List the prime numbers between
       40 and 50 _____

   (b) Write 40 as a product of its prime factors.

   _____

2 (a) Multiply 0.078 by 100 _____

   (b) Divide 3405 by 100 _____

3 (a) Round 207.8 to two significant figures.

   _____

   (b) Round 5.084 to 2 decimal places.

   _____

4 (a) Write the fraction $\frac{7}{20}$ as a decimal.

   _____

   (b) Write the decimal 0.8 as a fraction in its
       simplest form.

   _____

   (c) Write the fraction $\frac{3}{5}$ as a percentage.

   _____ %

5 Complete the multiplication square.

×	6	7	5	9
5				
9				
8				
7				

6 (a) How many chocolate bars priced at 65p
       could you buy with a £10 note?

   _____

   (b) Find the total cost of 4 books priced at
       £5.99 each. £ _____

   (c) At what time did a $1\frac{3}{4}$ hour film end if it
       started at 18:20? _____ : _____

7 (a) Subtract: 22.45 – 13.48

   _____

   (b) Divide: 11.6 ÷ 8

   _____

8 A calculator shows the result: [1.25]

   What would this mean in:

   (a) days and hours

   _____ day _____ hours

   (b) feet and inches?

   _____ foot _____ inches

9 (a) Complete this statement using **even**
       or **odd**:

       An even number minus an _____
       number gives an odd number.

   (b) When Freddie divided 124 by 8 his
       calculator showed the result: [15.5]

       Is this correct? _____

10 The nutrition information on a pack of two
   filled red peppers is:

	per 100 g
Fat	3.2 g
Carbohydrate	9.8 g
Fibre	1.6 g
Protein	2.4 g
Salt	0.5 g

   One filled red pepper has a mass of 150 g.

   (a) What would be the mass of carbohydrate
       in one filled red pepper?

       _____ g

   (b) What percentage of a filled red pepper is
       protein?

       _____ %

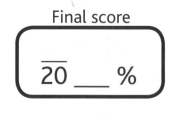
Mixed test 6

Questions 1 and 2 concern sequences generated from 2-digit starting numbers by applying a rule over and over again.

The rule is: Multiply the tens digit by 4 and then add the units digit. Stop when you get a single digit. For example:

$79 \longrightarrow 37 \longrightarrow 19 \longrightarrow 13 \longrightarrow \underline{7}$ (4 steps)

1 Write down the sequences for the following starting numbers:

(a) 23 _____

(b) 61 _____

2 Find two starting numbers between 30 and 40 which result in the single digit 9
_____ and _____

3 Solve the equations:

(a) $b + 8 = 4$ $b =$ _____

(b) $\frac{d}{3} = 15$ $d =$ _____

4 Complete the table of input and output values for this function machine.

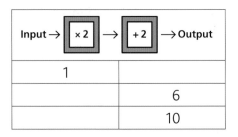

Input → ×2 → +2 → Output	
1	
	6
	10

5

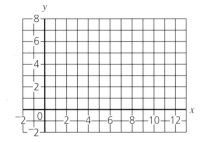

(a) On the grid, draw the lines $x = 8$ and $y = 5$

(b) Write the co-ordinates of the point where the lines $x = 8$ and $y = 5$ cross.

(_____ , _____)

(c) Draw the line with equation $y = x$.

6 For the cuboid, calculate

(a) the volume

_____ cm³

(b) the total surface area.

_____ cm²

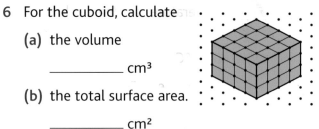

7 Complete the shape which is symmetrical about the line **m**.

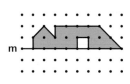

8 Calculate the sizes of angles d, e and f.

d _____ ° e _____ ° f _____ °

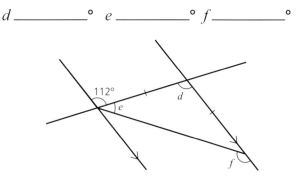

9 Terry's test scores are listed below.

8   9   8   6   7   8   8   6   7   7

(a) What is the range? _____

(b) What is the mode? _____

10 Fay has two ordinary dice.

Which letter on the scale below best represents the chance that the dice will show the same number when they are rolled at the same time?

_____

Mixed test 7

1 (a) What temperature is 8 degrees lower than 5 °C?

_____ °C

(b) What is the value of $^-4 + 7$? _____

2 (a) If these decimals are arranged in order of size, which will be in the middle?

_____

5.76   6.57   7.65   5.67   6.75

(b) Write down the number which is exactly half way between 53 and 81

_____

3 (a) Round 18.6 to two significant figures.

_____

(b) Round 8.99 to 1 decimal place. _____

4 (a) Complete these equivalent fractions:

$\dfrac{1}{3} = \dfrac{}{24}$

(b) Write the fraction $\dfrac{9}{24}$ in its simplest form (lowest terms).

_____

(c) Write the mixed fraction $1\dfrac{3}{4}$ as an improper fraction.

_____

5 Calculate:

(a) $3 + 5 \times 8 =$ _____

(b) $4 \times 7 - (3 + 2) =$ _____

6 (a) What is the sum of the first five multiples of 5, starting with 5?

_____

(b) Subtract 489 from 4000

_____

7 (a) Add: 49.3 + 4.93

_____

(b) Multiply: 458 × 17

_____

8 (a) Amelia calculated the total cost of her shopping and her calculator display showed the result 57.4

What was the cost of Amelia's shopping?

£ _____

(b) A calculator shows the result 4.8

What would this mean in hours and minutes?

_____ hours _____ minutes

9 Complete these statements using **even** or **odd**:

(a) An odd number times an _____ number gives an even number.

(b) An odd number minus an even number gives an _____ number.

10

Train	X	Y	Z
Meely	09:40	11:50	14:00
Nearly	10:20	12:30	14:40
Overly	12:45	14:45	17:10
Peely	15:42	17:48	20:09

(a) Which train travels from Meely to Peely in the fastest time?

_____

(b) How long does train Z take to travel from Nearly to Peely?

_____ hours _____ minutes

Mixed test 8

Questions 1 and 2 concern this pattern sequence.

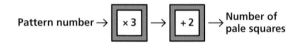

1   Complete this table of data.

Pattern number	1	2	3	4	5
Number of pale squares	5	8			
Number of dark squares	1	4			

2   Aidan suggests that you can use this flowchart to find the number of pale squares in any pattern number.

Pattern number → | × 3 | → | + 2 | → Number of pale squares

(a) Use Aidan's idea to calculate the number of pale squares in pattern 100

    _____

(b) A pattern in the sequence has 100 dark squares. How many pale squares does it have?

    _____

3   Walter has w marbles. Hettie has 5 more marbles than Walter and Sophie has twice as many marbles as Hettie. Write an expression, in terms of w, for the number of marbles that Sophie has.

    _____

4   Complete the table of input and output values for this function machine.

Input → | + 3 | → Output

1	
	8
	1

5   (a) On the grid, plot the input (x) and output (y) values for the machine in question 4

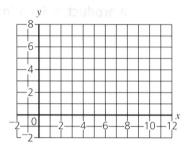

(b) Complete the equation of the line through these points.

    $y =$ _____

6   (a) Measure the length of line AB, giving your answer to the nearest millimetre.

    _____ mm

    A _____ B

(b) Write 2.07 kg in grams. _____ g

7   (a) Name the plane shape **S**. _____

(b) Mark all lines of symmetry on shape **S**.

8   Two angles of a triangle are 49° and 94°. What size is the third angle? _____°

9   The Carroll diagram was prepared by the members of Year 6

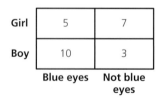

	Blue eyes	Not blue eyes
Girl	5	7
Boy	10	3

How many girls have blue eyes? _____

10 If a pupil is chosen at random, which letter on the scale best represents the likelihood that it will be a blue-eyed boy? _____

A   B   C   D   E   F

Impossible   Even chance   Certain

Mixed test 9

1 (a) List the prime numbers between
       60 and 80 _____

   (b) Write 42 as a product of its prime factors.

   _____

2 (a) Multiply 10.04 by 100 _____

   (b) Divide 26 080 by 1000 _____

3 (a) Round 44.49 to two significant figures.
       _____

   (b) Round 10.35 to 1 decimal place. _____

4 (a) Write the fraction $\frac{4}{5}$ as a decimal. _____

   (b) Write the decimal 0.12 as a fraction in its
       simplest form.

       _____

   (c) Write the fraction $\frac{11}{20}$ as a percentage.

       _____ %

5 Complete the multiplication square.

×	7	11	9	12
8				
6				
12				
11				

6 (a) How many books priced at £4.95 could
       you buy with a £50 note?

       _____

   (b) Find the total cost of 11 ice lollies priced
       at 99p each.

       £ _____

   (c) At what time did a $2\frac{1}{4}$ hour concert start
       if it ended at 22:10?

       _____ : _____

7 (a) Subtract: 41.08 – 4.18

       _____

   (b) Divide: 290.5 ÷ 7

       _____

8 A calculator shows the result: [2.05]

   What would this mean in:

   (a) hours and minutes

       _____ hours _____ minutes

   (b) kilograms and grams?

       _____ kilograms _____ grams

9 (a) Complete this statement using **even**
       or **odd**:

       The square of an odd number is an

       _____ number.

   (b) When Ariane multiplied 20.5 by 6 her
       calculator showed the result: [123]

       Is this correct? _____

10 The nutritional information on a 400 g
   chicken tikka masala is:

	per 400 g meal
Fat	9.6 g
Carbohydrate	52.8 g
Fibre	6.4 g
Protein	38.4 g
Salt	1.6 g

   (a) What would be the mass of carbohydrate
       in 100 grams of this meal?

       _____ g

   (b) What percentage of the meal is fat?

       _____ %

Mixed test 10

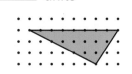
Questions 1 and 2 concern sequences generated from 2-digit starting numbers by applying a rule over and over again.

The rule is: Add the tens digit to the product of the digits. Stop when you get a single digit. For example:

$$45 \longrightarrow 24 \longrightarrow 10 \longrightarrow \underline{1} \qquad \text{(3 steps)}$$

1   Write down the sequences for the following starting numbers:

   (a) 53 _____

   (b) 99 _____

2   What is the smallest starting number to give the single digit result 9? _____

3   Solve the equations:

   (a) $a - 8 = 4$ $a =$ _____

   (b) $\frac{d}{5} = 10$ $d =$ _____

4   Complete the table of input and output values for this machine.

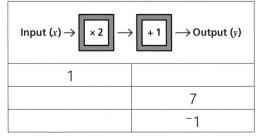

Input $(x) \to$ ×2 $\to$ +1 $\to$ Output (y)	
1	
	7
	-1

5   (a) On the grid, plot the (x, y) values for the function machine in question 4

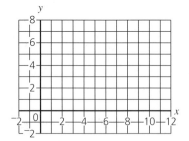

   (b) Draw the line through the points and complete the equation of the line.

   $y =$ _____

6   (a) Calculate the area of this triangle.
      _____ units²

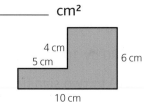

   (b) Calculate the area of the shape.
      _____ cm²

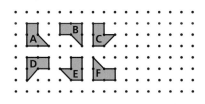

7   (a) Which shapes are congruent to **A**? _____

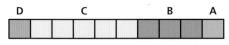

   (b) In a space on the grid draw a shape similar (but not congruent) to **A**.

8   (a) On the grid for question 5, plot the points (4, 4), (6, 4) and (6, 8) and join them to draw a triangle.

   (b) Rotate your triangle through 90° clockwise about the point (4, 4).

9   The diagram shows the proportions of exam grades awarded to 50 children.

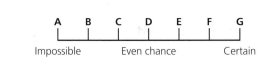

   (a) What percentage of the children achieved A grades? _____ %

   (b) What fraction, in its simplest form, achieved C grades?

   _____

10 On the scale below, which letter best represents the likelihood of getting a square number when a die is rolled?
   _____

   A   B   C   D   E   F   G
   | | | | | | |
   Impossible   Even chance   Certain